MW01620521

HOT RODS By ED "BIG DADDY" ROTH

Ed "Big Daddy" Roth & Tony Thacker

Motorbooks International
Publishers & Wholesalers

First published in 1995 by Motorbooks International Publishers & Wholesalers, PO Box 2, 729 Prospect Avenue, Osceola, WI 54020 USA

© Tony Thacker, 1995

All rights reserved. With the exception of quoting brief passages for the purposes of review no part of this publication may be reproduced without prior written permission from the Publisher

Motorbooks International is a certified trademark, registered with the United States Patent Office

The information in this book is true and complete to the best of our knowledge. All recommendations are made without any guarantee on the part of the author or Publisher, who also disclaim any liability incurred in connection with the use of this data or specific details

We recognize that some words, model names and designations, for example, mentioned herein are the property of the trademark holder. We use them for identification purposes only. This is not an official publication

Motorbooks International books are also available at discounts in bulk quantity for industrial or sales-promotional use. For details write to Special Sales Manager at the Publisher's address

Library of Congress Cataloging-in-Publication Data

Thacker, Tony
Hot rods by Ed "Big Daddy" Roth / Tony Thacker.
p. cm.
Includes index.
ISBN 0-87938-980-X
1. Automobiles—Customizing—California—History.
2. Automobile engineers—United States—Biography.
3. Roth, Ed. I. Title.
TL255.2.T49 1995
629.228—dc20 94-48214

On the front cover: Ed, at his finest hour, with the newly completed Beatnik Bandit photographed by Bud Lang. *Photo courtesy of Darrell Zipp.*

On the frontispiece: Hamming it up as usual in the offices of *Rod & Custom. Photo courtesy of Petersen Publishing.*

On the title page: Another Bud Lang shot of the Bandit. *Photo courtesy of Darrell Zipp.*

On the back cover: No, that's not a palm tree growing out of the engine of Ed's "Tweedy Pie", it just looks that way in this photo. Note Ed's attention to keeping the interior clean by removing his shoes and leaving them outside the car before posing.

Designed by: Kailay Yu

Printed in Hong Kong

THE
BOOK OF MORMON
Another Testament
of Jesus Christ

I'D LIKE TO THANK MY FATHER IN HEAVEN FOR THE HEALTH, STRENGTH AND REVELATION TO MAKE THESE THINGS POSSIBLE.

Contents

Introduction & Acknowledgments

Of course, I blame it all on designer Thom Taylor who, after seeing Ed's first book, "Confessions of a Rat Fink The Life and Times of Ed "Big Daddy" Roth," encouraged me to do a book concentrating on the vehicles that Ed had built using photographs other than the so often seen postcards.

Where possible, we've tried to use photographs from private collections that have rarely, if ever, been published. We'll admit that some of them are a tad below par but in many cases they were the only photographs available—we had no choice—and rather than leave them out we included them to complete the picture.

I had approached Ed some years ago but he declined my offer to work with him saying, "When it comes to publishing, I found out one must do it himself or be lost in political TURMOIL."

How right he was. We got into this project, or at least I did, but could find no common ground on which to work with Ed. It looked like the project would be stillborn until Ed's youngest son Dennis mediated and looking back, neither of us understands what the problem was. Nevertheless, I owe a big debt of gratitude to Ed, Dennis, and all Ed's family for their help and support with this book—I hope it is something they can be as proud of as I am.

Knowing Ed had suffered a devastating robbery in 1970 and acknowledging that he was always more interested in moving forward than looking back and therefore wasn't much of a "keeper" I knew at the outset that finding the photographs to illustrate this book would be one of, if not the, hardest part. Indeed, when freelancing for *Choppers Magazine* in the early 1970s, after Ed had sold it and it was owned by Hi-Torque Publishing I heard from then editor Chris Bunch that all the material purchased from Ed was floating in a flooded basement to be subsequently shoveled into the trash. Disasters notwithstanding, Ed's Mormon religion requires that all members of the church keep a record so at least some of his history, particularly the early, pre-business part was recorded.

Meanwhile, Thom had approached John Dianna and Bob D'Olivo at Petersen Publishing who kindly opened up the Petersen archives and provided access to some of the only construction shots taken of Ed's most active and perhaps exciting building period—the early 1960s. Those top-quality photographs were both the inspiration and impetus to continue—something I'm glad I did.

Next, I began interviewing people associated with Ed: George Barris, Bo Bertilsson, Gary Canning, Franco Costanza, Dirty Doug, George Goodrich, Joe Henning, Jim "Jake" Jacobs, Tom Kelly, Jim Linskey, Ed Newton, George "The Bushmaster" Schreiber, Larry Watson, Robert and Suzanne Williams, Dan Woods, Tom Vogele and the staff of *Street Rodder* Magazine including Rich Boyd, Eric Geisert and Jerry Weesner, and Fritz Voight, and raiding their personal photo collections.

Somebody who didn't work with or for Ed but who was a constant source of encouragement, background and the occasional rare photograph and needs thanks was Greg Sharp. Likewise, David Fetherston, Pete Millar, Dean Moon Jr., Dave Peters, Andy Southard Jr. and Bob Larivee who gave me access to his archives and turned me on to photographer Bob Hegge.

Bob had died in the early 1980s, but Bob Larivee seemed to think that Bob's nephew had saved the Hegge collection. A frantic call to Mrs. Hegge in Kansas revealed that her nephew had indeed saved the photographs. The stroke of luck was that her nephew, Doug Pizac, was an AP photographer living just a few miles

From left to right are: Dick Cook with the Orbitron, Dirty Doug with the Road Agent, and Ed with the Beatnik Bandit. ***Photo courtesy of Griff Borgeson***

from me in southern California. A call to Doug revealed that he had thankfully saved everything.

Another of many strokes of pure luck came when interviewing Tom Kelly who with his grandfather, "The Baron", was in business with Ed in the late 1950s. Kelly also had a good collection of photographs including a contact sheet bearing the name of photographer John Gregoire. Wondering aloud if he was still alive Kelly replied, "Of course, he's my uncle, he's chief photographer at the Jet Propulsion Lab right here in Los Angeles." A call to John revealed that he too had retained those shots taken in 1958.

Another photographer who was in the right place and the right time was Griff Borgeson who we tracked down to his home in France from whence came another package of priceless photographs.

At this point I thought I was doing pretty good but almost everything we had was black and white and the beauty of these books is that they can be, if you can find the photographs, all color. Sadly, nobody seemed to have any color. Then somehow I got the bright idea of appealing, through the various street rod magazines, for help. The response was unbelievable and some 200 calls later we had material from private collections, much of it in color, that we'd never have found had it not been for those magazines printing our appeal. We can't thank them enough, likewise the good folks who responded particularly: Jim Ballard, Fred Caldwell, Jim Handy, Roger Kilborn, Bob Lewinski, Mike Lowe, Lou Martineau, Mark Nelson, Terry Thompson, Doug Wright. We'd also like to thank both the National Automobile Museum, The William F. Harrah Foundation and the Southward Museum Trust Inc.

Last but not least we'd like to thank Darryl Zipp of Zipper Motors who for 12 years was Research and Development Manager at Revell and had the foresight to save a big bundle of photographs when all that stuff was trashed and Revell moved.

It's a long list of acknowledgements but we couldn't have done it without any of you so we thank you all including the good people at Motorbooks who agreed to publish. Everybody was giving of their time and materials and we thank them all for that. If I've inadvertantly missed somebody, please accept my sincerest apologies.

What follows is Ed's account of what happened, not mine, I wasn't there, however, Ed's the first to admit that his memory's not exactly spot on, especially when it comes to dates. So unfortunately for those retentive readers we have not attempted to attach specific build dates to vehicles because looking at the photographs reveals that a lot of stuff was going on at once. For example, construction of the Surfite overlapped the construction of both the Orbitron and the Wishbone and the photographs were taken years apart. Besides, what would you use as a benchmark other than memory. Certainly not the magazines: They would shoot material and put it on file until time and space permitted publication. Besides, magazine production lead times of up to three months throws that dating medium outta the window.

We make no apologies for not trying to be specific about exactly when each vehicle was built. It matters not when they were built but that they were built. Meanwhile, to get a viewpoint other than Ed's, I interviewed many of the people who worked for him or with him over the years. Read on and Ed and I hope you enjoy the fruits of his labors

Tony Thacker
Huntington Beach, California, 1994

Y'know, it Gets in Your Blood

Seems like I always been drawin' or buildin' stuff. When I first went to kindergarten, in 1937, I remember not bein' able to understand what the teacher was talkin' about. I had this slight problem, ya see I couldn't understand English. In them days nobody cared about bi-linguals & I was one. My folks'd drifted in from Germany in '28 & so that's all they wuz able to talk. So natch, all me & my brother knew when we hit school was German. I guess it was sorta cool though, 'cause whenever some bigger kids got in our face I could tell my bro (In German) to tell my mom & she would take care of business with that big broom o' hers.

Then came WWII & things got outta hand 'cause whenever some kids would come to pants me (take my pants off an run 'em up the flagpole) for bein' a Nazi I didn't have any friends to help me out at school. Most o' my relatives in Germany wuz Nazis, but my dad was true blue American & raised me & my bro the same way. If anyone came with the swastika, or Nazi stuff of any kind, he'd tell 'em to take a hike. Regardless of all that, I was like any other kid in school full of adventuresome mischief & looked at the war as sort of an escape into my own little world. Man, I'd watch the newsreels of how the Japs & Germans had all this real slick stuff like rockets & Kamikaze pilots & my pencil & brain got wound up in some real flights of fantasy.

I mean, who cared about what the teacher wuz tryin' to put down about math or chemistry. That was all nonsense (still sorta is) but this great big world out there was just waitin' for me to come & rescue it. My two-holed notebook became a battle ground. Jap, German & American planes & soldiers battled across it's pages with unequaled bravery. Bodies & blood & wrecked airplanes became an extension of the gory newsreels we saw at the movies on weekends (Yup! It was before "The Tube" came along).

After school we would make machine guns & bombs out of plywood & coffee cans & lay in the fields in foxholes taking turns at being the bad guys. I think my mom even made some fake Red Cross bandages for our team. In '45 the war ended & so did my artistry. It wasn't exactly the peace that did it but I was 13 & I was gettin' primed for my driver's license.

In them days, 14 was drivin' age! Whoopee! Course in '46 L.A. was still a young city & Bell was on the outskirts. Still some farmin' goin' on but whenever we went somewhere, my favorite aunt's for example, my dad'd let me drive one way in his '34 Chevy sedan. No driver's training. Just me & dad. Anyway, I was buggin' pops to let me get this '34 Ford. Boy was he nervous. Man, I use'ta dream about it at nights, but he put the nix on it when he found out there wasn't a pink slip for it. Parents're funny that way. So I saved my gold for this really cherry '34 five-window to drive to school. I paid $350 for it & my dad almost became unglued. I was king o' the hill. The best there was at Bell High School (near South Central L.A.) in them years—I remember pickin' up Sally (my first main squeeze) & her girlfriend Dianeto to go to school.

All us guys had rods. Hot rods to be more exact 'cause they boiled over a lot. That came from the big mills we was always stuffin' under the hoods. We hadda get what we could from the junk yards since "Uncle Sam" had collected the really hot iron for the war effort. Our cars were the best of the worst.

In those early years there was no such thing as rod or customs. They were all labeled hot rods. Even the lead sleds. As '47 & '48 came along ya'd see more & more guys takin' the later fat fendered Fords & Chevs & fillin' in the nose & deck emblems with lead & lowering the back ends. Then choppin' the tops & so the term lead sled or "sleds" became popular. Sleds weren't fast, just looked cool. They were more for "cruisers" out scoutin' for the babes. Pontiacs or Mopars were nerd cars at the time & ya could pick 'em up for a song. La dee dah!

Usually the really hot combo was a set of '32 or '29 rails with a flathead V8 engine out of a later, '39 to '46,

Ford. Sure there was plenty 'o guys pickin' up on not so groovy stuff like '35 Plymouth coupes (Yech!) or '39 Pontiac coupes, but in them days it was strictly a No! No! Then there was the street races at night where we found out who had the fastest machine. This was not my bag. Seems everytime I pegged my ride (went fast) I'd break the engine & cost bucks, so I became a cruiser at an early age. I still drive most o' my stuff at a snail's pace.

I think that's what made guys build faster cars. They wuz into this speed thing. Y'know, it gets in yer blood. Make a buck n' spend it on speed equipment. In downtown Bell there was this guy named Roy Richter & he was a real cool flattopped racer. (Hairstyles in the 40's was flat tops with a Duck Butt in back.) Somehow he had a hobby o' makin' some kinda oil well parts. The name of his company was Cragar, from the name Crane Gartts who started the company, but Roy made some really nifty stuff that made us go faster.

Below: The summer of 1950 saw Ed's 1950 Chevy coupe sporting purple paint and fake Caddy hubcaps bolted to the rims for security. Its rear fenders were molded into the body on one side with lead and on the other side Ed experimented with a new material called fiberglass. Too slow, the car was sold to the Bell chief of police. Above left: Ed's first hot rod, a '33 3-window, painted in the driveway with his mom's vacuum cleaner. ***Photos courtesy of Ed.***

F'rinstance whenever ya wanna go faster ya gotta get more gas to the cylinders. So Roy sold this really slick intake manifold that had two 97's on it. Then most of the guys milled their heads. About 100 thousandths shaved from the heads would increase the compression ratio from 7:1 to 10:1. The extra gas would work better if the heads were a higher compression ratio. Roy sold finned-aluminum high-compression heads with names like Edelbrock, Eddie Edmunds, Offenhauser, Iskenderian, etc. Then stroking the crank & putting high-compression pistons & specially ground camshafts: Potvin, Isky, Crower, etc., would again increase the horsepower of those poor overworked flathead Ford engines. In them days Chevys were for old people. Fords wuz all we had 'til '55. Sure, there was a coupla nerds who bought Chevys but they were leadsleds strictly for cruisin'.

Well some o' these guys got real serious about racin' and we'd go to the drive-in: Bell Clock, Hula Hut, Stans, etc., & the hot dudes would "choose off" one another for a race. Well this was no simple situation. We'd say, "What ya got under the hood?" Terms like "full race" & "1/4 race" were common expressions of what their engines contained. "Strokers" were the feared breed of engine. If anyone had a full-race stroker he had the whole enchilada. The whole nine yards. Can ya dig this? We could tell by the sound of the motor what it had in it. Sounds far out, but true. Then we'd all pile into our cars & go to Slauson Avenue (before the freeways) & race a quarter mile.

The more serious racers raced for pink slips. I've never personally seen anyone collect off'n this kinda deal, but the guys were always talkin' about it. Ya gotta remember that all o' this was hush hush 'cause we wanted to have three or four races before the heat came screamin' down Slauson with their sirens & lights blastin' away. There was about three to four hunnert of us out there. Some would go hide in the cemetery & some o' us would get in our cars & silently cruise back to the Clock. Later on the cops gave tickets just fer bein' in the area.

I remember, we had this real nurdy cop called "Dirty Ernie" & he'd write a ticket for smitties (loud mufflers) or fishtailin' wheelies. No sense o' humor! Then the cops couldn't figure out if it was legal or illegal to be fenderless. Sometimes, depending on the kid, one guy'd get stuck with a no fender ticket & sometimes not. The judge's kid always seemed to get away with no fenders. Wonder why?

Then there wuz guys like Gaylord Lunney who goes by "Gaylord" in the rod mags with his pickup bed covers & liners. Well, Gaylord was the best example of American enterprise that I know of. Ya see Gaylord had two things he was good at: women & cars. He had 'em both. He was the head jock at Bell. Later on, when he went into the upholstery business, he hung all o' his tickets to the wall. "Count 'em," he'd dare ya! There was too many, but we took his word for over 200 of 'em. When he was 14 he got this old wreck like the rest of us but instead o' goin' to Tijuana for seats he did 'em himself

& it wasn't bad. He painted it & sold it for a fortune & he took the bread & bought another wreck & did this about six times & ends up with this beautiful, brand new '48 Caddy all decked out with a "Carson-type" top on it that he made himself. Girls? Ya can imagine.

In those days it was just customary fer girls to take off their shoes whenever they stepped into a dude's ride. After spendin' umpteen hours workin' on the insides & carefully cleanin' the tuck n' roll upholstery that we got in Tijuana, wasn't nobody gonna destroy any piece of it if we had anything to say about it! But lookee here, I wasn't the only one. All of us did it. In them days there wasn't any "daddy's cars" y'know, we hadda make our own cars. Build 'em from scratch.

So I says to myself, "If Gaylord can do it, so can I." That's why I always have tried to do my own stuff. I say tried. I never could get a handle on leather. I've always hadda depend on guys like Martinez, Perez, & "Downtown Willy" to get that done for me. So in '48 I got me a '39 Chev coupe & leaded in the rear fenders. I was only doin' what came naturally. That is what every one else was doin', so I did it too. Painted her purple. Split manifold & fake Caddy hubcaps made me the king of the Bell roads. But it was definitely slow so I switched back to a more conventional '32 three-window. It had moxie. It was already outfitted with 16-inch skins & solid wheels off a later model Ford. One neat thing about Ford was

Below left: It was love at first sight but Ed soon removed the hump from the grille as can be seen in this 1953 photograph. Below right: Earning a living pinstriping in the street. Left: To replace the Chevy, Ed bought this raked and mildly rodded '32 3-window seen here in January 1951. By May he'd removed the bumpers and fenders, channeled the body and almost given his dad a heart attack. *Photos courtesy of Ed.*

that the wheels & motors & running gear from the later models fit into the earlier models. Even the hydraulic braking systems.

I worked for the money as a box boy, newspaper routes, truck washer, anything to make a buck 'cause my dad was a struggling dude tryin' to feed the family plus he believed as I do, if ya want somethin' bad enuff go to work & save up for it.

This '32 was stock bodied when I got it in '49 but I soon hammered & slammed it. I drove it for about a year & then took the fenders off & channeled the body over the frame.That was the closest that my dad ever came to a heart attack the day I did that.

Before I get too wrapped up in details I gotta mention that my folks always supplied a place for me to work. And though my pops never agreed to me channeling my '32 coupe, he helped me all he could. All my friends had dad's of the same caliber & some of the dads even helped us modify some of our cars.

Gettin' back to all of this confusion of what we were drivin', *Hot Rod Magazine* started calling some of our cars hot rods & some of 'em customs. I had both. I had the '39 Chevy sled & I had the Deuce coupe. I preferred the coupe but drove it like a sled. However there wuz some guys in the crowd that had some better sources o' bread than I had. They were able to feed those engines 'cause as the saying goes, "To go fast ya gotta have cubic bucks." These're the guys that forgot all about the height of the license plates & the noisy mufflers & just went fer broke.

I know what you're thinkin', "What has all o' this got to do with your cars "Large Father?" Well I mention all o' this 'cause it tells ya exactly where I'm comin from & what kinda guys I was hangin' out with. Y'see if it wasn't for guys like Gene Thurman, Mickey Thompson, Dean Moon, Gaylord, & Fritz Voigt, I'd never have been able to get into this car buildin' mode that I got into.

My graduation from Bell High was average. Basic college was the same. I had majored in math at East Los Angeles Junior College but my best work was done after school workin' on my cars or helpin' my buddies or hangin' out at the Clock Drive-In in Bell, eatin' burgers & romancin' my first wife Sally.

In '51 I put my coupe up on blocks in my folks' garage & joined the Air Force. As soon as they found out I had a basic idea of mechanics, they sent me to bombsight school. Then to South Carolina & then Africa fixin' bombsights, workin' in photo labs (there is a connection y'know!). At the time I had me a '46 Ford & then a Henry J. It was love at first sight when I saw that '51 model. Hadda have one. I've got this unusual knack of pickin' "winners" at a glance. The first thing I did was to take the hump outta the middle of the grille.

I hadda dump the 'J' 'cause I was gettin' back into civilian life & couldn't handle the payments. That's when I latched onto this '48 Ford. It was cheap & all I did was paint it red (my fave color fer sure) & whenever I wasn't "decorating" some dudes ride I was paintin' & stripin' my '48.

In '55 those squares at Chevrolet had the nerve to make the '55 Chevy with not only this real cool body but it had this big muscle under the hood in the form of an overhead-valve V8 engine courtesy of Zora Arkus Duntov. This single car wiped out all the coolness that Ford had developed since '32. I mean this '55 was a wicked mosheen. Fast & lookin' good! Guess what? The coupe I had in my garage was history. I sold it. I was lucky to get three hunnert bucks for it.

In '56 I got a big surprise when I went to look at the new model cars. Chevy had another winner & Ford was runnin' far behind. They'd upgraded the smooth little Duntov V8 & improved the looks. I was almost ready to go take the plunge but financially hadda wait til '57 to get me a newer car. The '48 Ford with the head on the back drew too many complaints from people & besides I was makin' out pretty good & I deserved a new car so I bought this brand new '56 Ford pickup to advertise flames & car names. Up to '55 pickup trucks were very utilitarian. Y'know, haul the wheat to town type of thing. Ford broke thru that design barrier & made a truly great customizable pickup in '55-'56. I flamed mine & got a lotta customers off the street that wanted "flames". I sold it to a guy that had to have it with a big roll o' loot

The other '55 Ford truck I had was one I bought from Gene Moore. He was a Bell High buddy & he'd built this slick '55 & needed bread for another project so he bailed out of it for 1,200 bucks. I towed some of my show cars but it was too low & too nice so I sold it & got a hearse to tow with.

I spent three years at Sears thanks to my mom-in-law who got me in as a window display decorator from '55 to '58. It was about that time that Von Dutch was makin' his debut as a pinstriper. He was puttin' all these funny marks on cars & people was all gettin' it done to their cars. All's he was doin' wuz covering up gliches in paint jobs, like grinder marks or rock marks. He called 'em "Chicken Scratches". All's Dutch was doin' was takin' the round grinder marks that came thru the lacquer paint (paint in them days'd shrink forever) & then he'd make all those marks be a design. He had it sorta made cause his dad was a sign painter so he knew how to use a brush.

Well this wuz right up my alley. I got my first taste of paintin' signs in the Air Force in Africa. Y'know how it is with 80 guys in the middle o' the desert that're supposed to be doin' GI business. We were watchin' for UFO's & Russian aircraft. Trackin' 'em on this sophisticated radar equipment. At nights & weekends it was the California guys against the rest of the troops. We talked about cars & they talked about babes & baseball. Part of the car talk was about car clubs & I had some farout drawing & my name on my duffel bag. So at nights the other guys would have me do their bags. No big

Unable to keep up the payments after leaving the Air Force, Ed sold the J and bought this '48 Ford which in his spare time he painted red and decorated with pinstripes and that papier mache head to advertize his fledgling business. ***Photos courtesy of Tom Kelly.***

Working out of Stan the Man's shop in South Gate, Ed's business was taking off and he used this radically flamed '56 F-100 with an airbrushed mural on the tonneau cover to advertize his talents—"any car $4 an up." ***Photos courtesy of Tom Kelly (above), Greg Sharp (top).***

thing. Findin' the French paint that would stick to the canvas is another story. I painted signs on everything from tent plaques to my own barbershop—a plane engine crate tipped on end. Cars was just another medium. So me & about five other guys: Baron, Kelly, Slimbo, Jeffries & Watson, got on the band wagon. It helped me to support my family. Later, I quit Sears & pinstriped full time.

I'd talked this body shop owner in South Gate (Stan the man) to lease me one of his paint booths. It was a very busy street & I needed something to direct attention into the shop. Being the ham that I am I get the idea of puttin' a paper mache head of myself on the back of the '48 to attract attention & increase business. Sally (mi numero uno mamacita) figured I was nuts. Y'know what? I had so much business I never got to "practice" again. In fact I made enuff bucks to buy me a brand new '57 Bel Air which I quickly converted to the "Roth Look" with the top lined off in the new silver tape and what else? "Roth flames."

The Baron, Roth & Kelly

Kelly & his grandfather, Bud "The Baron" Crozier, were goin' around pinstriping cars at car lots. The Baron, an ex Studebaker employee, did stripes on the Studebaker assembly line & while he did the pinstriping at the car lots he got Kelly to do desert scenes on the gas filler lids.

One day in 1957 The Baron sees me & axes me if I'd be interested in joining forces & I says, "Yes! Indeed!" So we rented a shop eight blocks north of the old Barris Kustom City & started doing our "thang". Kelly had a '49 Chev & the Baron drove a Studebaker Skylark & I had my '57 Chevy Bel Air—one o' the sharpest off-the-showroom-floor mosheens I ever bought. Hot Dawg! It looked & ran like a million bucks.

Flames were just makin' the scene. About a mile west of our shop was this dude named "Earl". He'd done some fantastic flames on a '56 pickup truck that was hot stuff at the drive ins & I took a trip to see Earl. I had the idea of flamin' my name on the side of my '57 for advertisement & coolness. Earl was willin' to show me the basics. He sketched off the flames with chalk but Earl had this booze habit that made him very undependable & I finished the job.

I still had the hots for a rod but didn't want to fix up the '48 'cause my kids was gettin' sucker juice on the seats so the '48 became the first "beater" that I had & on the side I fixed up this cool '30 Olds-powered Ford two-door sedan. I called it "Little Jewel" 'cause guys were startin' to call their cars neat names 'cause Duane Steck named his '54 Chev custom "Moonglow" & so we all copied Duane. Jewel was red & with that big overhead V8 Olds engine & tranny under the hood it would scoot right along. I dumped it in the front and started taking it to cars shows but eventually I sold it to get bucks for the Outlaw. I needed fiberglas & tires etc. Goodbye "Little Jewel".

Goin' to car shows was never intentional on my part. Those first car shows, circa '56, held at the local football fields at Huntington Park High & parking lots behind Penney's were a family get together sorta like a cruise with no rules, no admission, just a good ol' time.

The devious thing in each of us car owner's brains was, "How'm I gonna one-up these turkeys the next go round?" This became the driving force of strippin' down our cars & doing some stupid thing like chrome plating the transmissions.The Collins Boys were upholsterers from Whittier and they chromed their tranny. So I tear down the '39 three-speed box, take all the gears out & take it to Model Plating along with the oil pan. It's so ridiculous I still can't believe I did it. But at the next get together I had the edge and the start of a chrome undercarriage By the time '57 rolled around, everyone had chromed trannys. Now What? Only solution was to build a ground-up car

We did a lot of scallop jobs in those daze. We worked over there for about three years. The Baron was collectin' social security & wasn't interested in buying a shop & I was so we split company in 1960. The Baron passed away soon afterward & Kelly is still doin' signs on stuff like Miller Beer trucks & all sorts of graphic designs along with his sons in Paramount. Kelly wasn't ever too hot on hot rodding. His whole life revolves around his wife, kids & home. Hooray for you Tom!

All this time I was carvin' away on the "Outlaw". The Baron use ta get upset a lot 'cause I'd take every spare minute to go to junkyards & work on the design. Ya gotta remember that this was a trendsetting situation. I'd never seen anyone do such an outrageous design, 'specially in fiberglas. I mean we were busy all the time. We even got a fourth pinstriper, "Slimbo" to come help out. He was one of Von Dutch's students & all I can remember about "Slimbo" is that he chewed on this big cigar that was never lit & he moved to Arizona. I remember askin' Baron about buying a piece of property & buildin' a shop & he said, "Ain't interested 'cause I don't have long to live". I didn't understand him at the time but I do now! Soon I had saved up enuff to buy a 27-foot lot in Maywood (next to Bell) & got Phil Lang to build me this tiny one-car garage.

Below: The Baron, Roth & Kelly striped and scalloped dozens of customs like this '57 Chevy. Left top: The Baron applies a name to the fender of Ed's flamed '57. For this they charged a mere dollar a side. Left bottom: Ed's brand new '57 Chevy with not only a wild flame job but also a chrome tape striping job on the roof. ***Photos courtesy of Andy Southard Jr. (color), John Gregoire (B&W).***

Left: According to Ed, he and The Baron would get the customers to guess which had the longest nose—you choose. Above: A typical Saturday in September 1958 at the Crazy Painter's shop at 9001 Atlantic Boulevard, Southgate. The purple Chevy on the left belonged to a young Kelly. ***Photos courtesy of John Gregoire (left and right), Andy Southard Jr. (above).***

Painting
LO. 64363
"THE CRAZY PAINTERS"
BARON and ROTH
50 years experience pinstriping
9001 ATLANTIC Southgate Cal
9001
SCALLOP
Crazy Painting
AUTOCADE
BARON ROTH

Below: Little Jewel, an Olds-powered '30 Model A Tudor was Ed's first forray into the show car scene. Wally Jordan offered to paint it for free as long as Ed didn't watch. Later, he confessed that he used plain enamel heated on a small stove until it was runny enough to spray without the use of thinner. Ed sold the Jewel to Lee Rhodes of the Long Beach Renegades in late '58. *Photos courtesy of Tom Kelly (top left), Greg Sharp (top right), and Franco Costanza.* Right: Tom Kelly in 1958. *Photo courtesy of John Gregoire.*

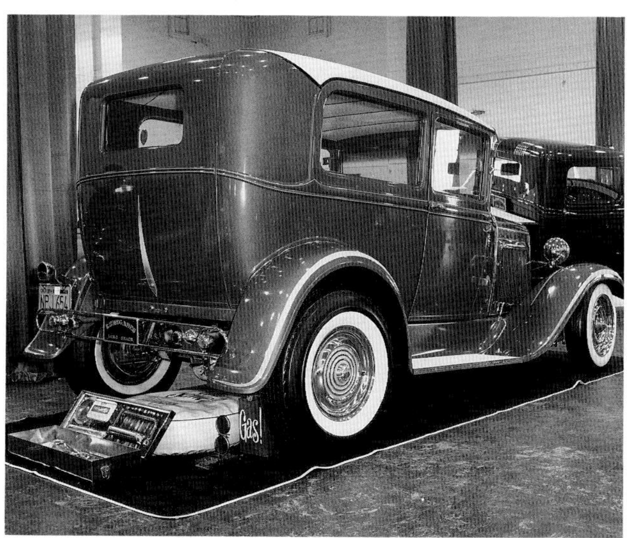

Tom Kelly

My grandfather and myself were striping on the car lots over in Maywood and Southgate. Ed was in Southgate doing striping and stuff and they decided to merge roughly around 1955. That first shop was called "The Crazy Painters" and it was at 9001 Atlantic Boulevard in Southgate.

My grandfather used to stripe for Studebaker on the old wagons and buggies, back in the early days, then he striped for Ford for 25 years on the production line back east. They gave him the nickname "The Baron" back in the Ford plant, that's what they used to call him, "The Baron of Striping." Then he moved out here.

The idea for doing T-shirts kinda came from Roth, he started doing some of his own and his friends' and we started from there and we were doing roughly 35 or 40 a day and we used to ship them all over the United States.

Most of what we were doing was the shirts but we were also doing scallops and pinstriping and we were doing a lot of names on cars—that type of thing. We charged a dollar a side. A scallop job was anywhere from $75 to $150, flames were maybe $200. We did a lot of them. We did 'em from 100 miles around.

I was painting shirts most of the time, Ed and my grandfather were running the business, but Ed would do a lot of the scallops and stuff. I was supposed to be doing the dashes—all the weird pictures, cartoons, flowers, etc. Ed did most of his stuff at night.

Ed's a hustler, he likes to do a lot of weird things, he's got a lot of imagination but we basically all got along pretty well. However, my grandfather and Ed had a little conflict going because we used to be open from 10 o'clock in the morning to 10 o'clock at night seven days a week. Our motto was, "Open 10 to 10, see you then." That corny kinda thing. They were partners, whatever was done they split but after my grandfather would go home Ed would come back and take appointments at 11 o'clock on his own. I guess that caused the split. That was around 1959.

At that time I had that little ol' '51 Chevy fastback, my grandfather drove a '58 Studebaker Blackhawk, and Ed had a '57 Chevy but he also had the "Little Jewel," a red '31 Ford Tudor.

The Baron was 67 when he died in 1962 and he'd always said to me, "When I die, you go to work the next day. Don't bother with me. When I'm gone you just get your fanny to work the next morning. Do that for me." That's what I did. I've been doing this since I was 13. ❐

Outlaw

In Africa I had got this fantastic idea for a fiberglas car when I saw a picture of Henry Ford beatin' the trunk of one o' his new '41 Fords with a sledge hammer & it wouldn't dent. Ya could'a knocked me over with a feather. It was also very cheap! It could also be done by people with little or no talent and I had both. It seemed too far out for my brain so I just dismissed it 'til I saw that *Life* article. In '57 I started playin' with "glas". I got some of the gooiest messes ya'd ever wanna see. My pants are always ruined by the end of each day, but in them days I'd have to throw 'em away each day. Shoes wuz good for about 4 days before I'd throw a coat of black paint on 'em.

First I made the frame which was your basic '29 Ford rails & fitted this junk Caddy engine into (junk but ran good). I knew fiberglas existed but couldn't get anyone to help me (except Dirty Doug later on) so I was gonna make me a body outta wood like the Shadoff Special guys'd done. But it was too complicated & besides, wood & me don't jive! So I went to the local lumber yard & got some casting plaster (which is gross 'cause it dries so quick) but it was cheap & better'n wood.

Makin' the buck was no problem. Guys in Detroit was usin' clay since the early 20's but clay was a buck a pound. Forget it! Plaster was a buck for 100 pounds. I used that! Then I covered the plaster with this messy, ooey, gooey stuff. I mean, like, it just ran into this big pile of mush on the asphalt. It was devastating. I couldn't ask anyone for advice 'cause no one knew. But I did it over & over 'til I got it right. When the glas cured, I knocked the plaster out from the backside and that became my "Outlaw" mold. (Robert Williams has that still now.) I made two bodies & sold one to some unsuspecting dude down the street. I only made one grille shell though so tough toodies!

I had few tools in them days. I had some basics like a crescent wrench & a hacksaw & a coupla files. So I wired, with bailing wire, all the stuff I wanted welded & took it to a trailer place down the street where this dude named Clarence Bell welded the stuff together. Then I took it home & filed all the weld marks down & took it to the chrome shop.

I gotta give Fritz Voigt, a local mechanic & sidekick of Mickey Thompson, a lotta credit for help in advice & parts. Meanwhile, I sold "Little Jewel" to get money for the chrome for the Outlaw.

As I worked on Outlaw I was plannin' a tow rig & trailer. Enter Dick "The hammer" Cook. Dick could solve any problem with a hammer. Dick had put an Olds Tri-power into a '40 Chevy & I bought it from him. It hauled bananas! It

Above: Ed gets to work in the street outside his new shop at 4616 Slauson Avenue, Maywood, with the plaster of Paris. Note the new sign (left) made by Bill Glanz, the shirt homemade by Ed and the garage door painted by Bell High School grad Gary "Snick" Conner. Below right: Ed still had his '57 but by now it sported his new address above the side trim. Ed knew then it paid to advertize. *Photos courtesy of Petersen Publishing (above) and Griff Borgeson (left and right).*

Above: The only set of female molds still exist in the care of Robert Williams. Bottom: Not the shot so often seen but another from the same day with Ed hanging onto his mom-in-law's revolutionary sword. The tow bar enabled Ed to wire together parts and tow it off to Clarence Bell who welded them up. *Photos courtesy of Petersen (top and middle) and Griff Borgeson (bottom).*

was perfect for a tow car. A little paint & a coupla signs. It was even big enuff (A sedan yet!) to sleep in. The nifty part of the '40 was the gear shift. Dick had made the emergency brake handle into a gear selector & when people offered to drive it & they figured out how to start it (another story altogether) they couldn't figure out how to make the thing go". Tough!

I can remember cruisin' down Hollywood Boulevard one day racin' this new '69 Chevy & leavin' 'em behind. At the next stoplight I glanced over to see that it was Ricky Nelson. Whenever I got to the shows it was neat to unhitch the trailer & go street racin'. The only un-stock thing about the exterior was I had my name on the door in small letters.

The trailer was a "Keep it simple" model courtesy Ron Aquirre's dad, Louie. It had two tires (one axle) & was the open air variety. Ron built "Xsonic", a bubble-topped Corvette with the first hydraulics courtesy Air Force salvage store. Me & Ron toured the east together in those years.We upset a lotta people with our outrageous cars.

Well you can imagine the excitement when I first showed the "Outlaw" (I named it the "Excaliber" at first 'cause I used my mom-in-law's family Revolutionary War sword, but no one could pronounce the name so I changed it to "The Outlaw" in about 1960). All the guys had been aced out with one grand swoop. I had everything chrome plated. I took trophy after trophy. Guys were throwin' their hands in the air. How was they gonna beat me? It was really unfair for me to compete & after I got aced out at the Oakland Roadster show (Around 1960) for usin' the new nylon lock nuts and not havin' cotter keys in my ball joints I never again tried to compete for trophies. They are spiritual & I carry them in my heart. I figure my stuff is so whacky it can't be judged anyway. I traded for a T-shirt booth or went for appearance money.

I remember jammin' thru Kansas with my open trailer & lookin' thru the rear view mirror to see if the

Right: The first outing to a show in the JC Penney parking lot in Huntington Park. Note the tonneau cover for lack of an interior and the solid front wheels. Bottom: Canning Hardware provided the aluminum sheeting for the gullwing top which unfortunately blew off somewhere in Kansas. *Photos courtesy of Dirty Doug (color) and Greg Sharp (B&W).*

Ed makes himself at home in this '57 S&S-bodied Cad hearse, the first of many. In this particular one he was stopped and searched for illegal immigrants on the way back to L.A. from San Diego. ***Photo courtesy of Ed.***

Outlaw was doin OK & I see the top had blown off in the crosswind. I didn't go back to find it 'cause it woulda been destroyed anyway. The poor farmer that found it musta thought he had a piece off a flyin' saucer.

Another time Ron was cruisin' & had two flats in a row so we put tires on his back axle (he had a 4-wheel trailer) & the front axle had no tires. Nooo problem! We boogied to the nearest Sears store for new ones & kept goin'.

Later on as the show circuit got busier I bought me a '57 S&S-bodied Cad hearse that I could sleep in. I had the back all decked out with TV & bed & food & stove. Y'know the real important stuff. Trouble was whenever I'd park somewhere to rest the cops'd swoop down on me & surround me. They'd knock on the back door & take my license & run a "Make" on it & hand it back to me. I got so use to it that I'd have the license by the door & hand it to 'em & go back to sleep til they got their jollies & then they'd give it back to me with this kind word of advice, "There's a rest area down the road please go there & rest." There was always a rest area down the road someplace.

Finally, I got wise to why they were snoopin'! My trailers!! They were so wild with big blatant signs & pinstriping in clown colors. So I got me a cherry '60 hearse & painted my trailer black to match & painted "Chapel of Memories" on it & never got bugged by a "Smokey Bear" again.

I guess the best part of the Outlaw was when Revell Model Company called me & wanted to make a model of the car. It was thrilling to think that Outlaw was chosen to compete with the likes of the AMT "Ala Kart" kit. By the by, that's when Revell tagged me with the "Big Daddy" moniker.

Most of the guys I knew had nick names. Kelly

Above: New dragster-style wire wheels now grace the front of the Outlaw along with the first interior. Right: Ed celebrates the introduction of Revell's Outlaw model and the impending royalties. Far right: The infamous revolutionary sword that gave "Excaliber" its first short-lived name. ***Photos courtesy of Ed (above), Darryl Zipp (right), and Greg Sharp (far right).***

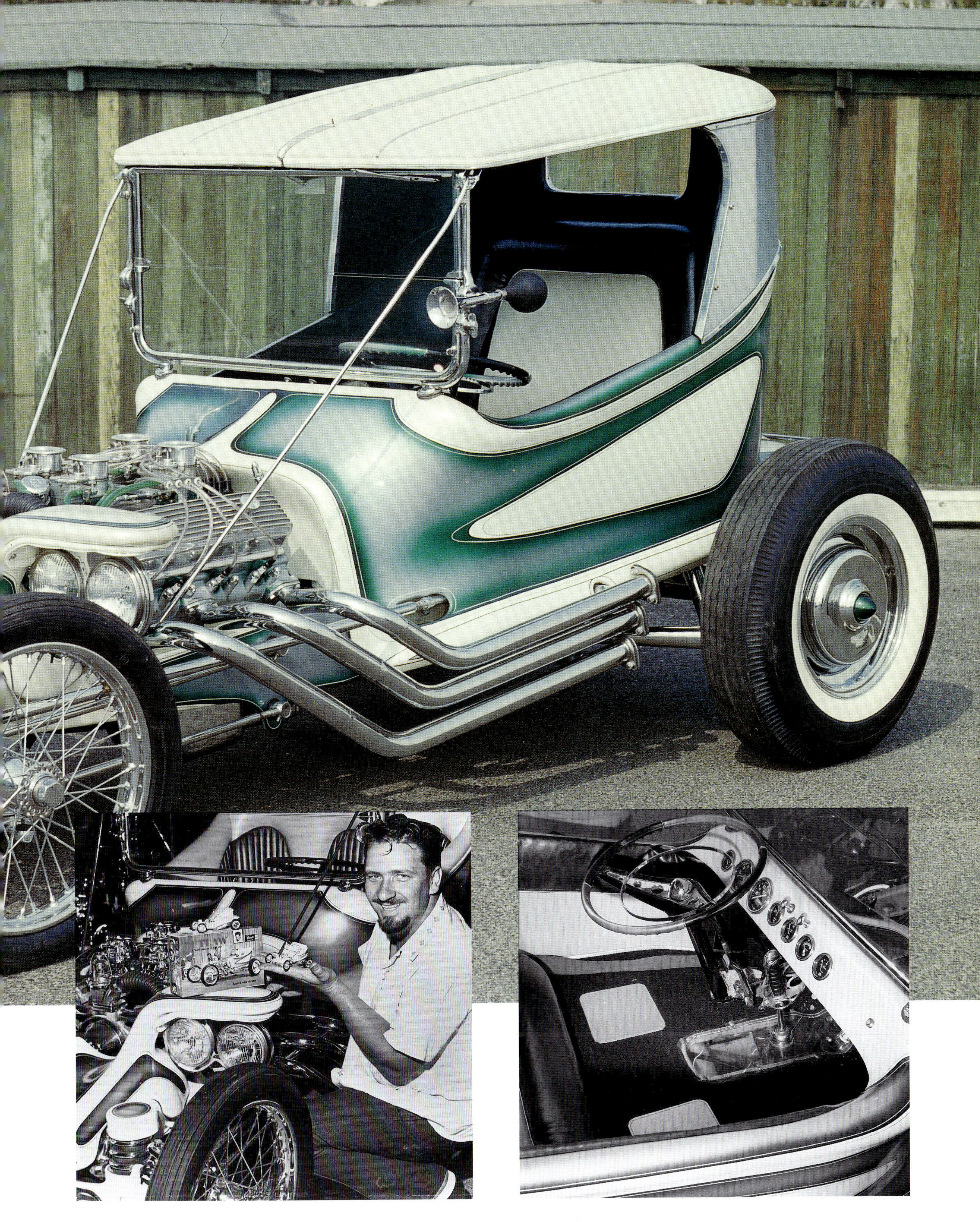

Above: The Outlaw at Darryl Starbird's Wichita show in the early 1960s. In the background: Starbird's Predicta; Dave Stuckey's '32 Tudor; and Ray Farhner's Eclipse. The little tape recorder (left) was used at shows. ***Photos courtesy of Roger Kilborn (above) Doug Pizac and the Bob Hegge collection (left).***

never went for it, neither did Watson or Jeffries. I always thought they would have made out better if they did. I don't know where Von Dutch or Slimbo or even The Baron picked up on their aliases but when Revell started making the Outlaw as a model kit Henry Blankfort called me into his office & said, "Your name needs a boost for the kits." I agreed but never was sharp enuff to pickup on a groovy "nom de plume". I'd always been called "Big Ed" at Bell High. But Revell figured that I oughta go with the "Dutch" movement & call myself something like "Spider" or "Roach" or sumpthin' cornball like that. It didn't appeal to me a bit. At about this same time there was this hippy movement in LA where guys'd let their beards grow & recite poetry at these coffee houses. The head stud muffins for these places were dubbed "Big

Below: After a rough life in show biz, Doug repainted the Outlaw metalflake green as it was seen in the Cars of Stars museum. Bottom: The only other Outlaw body seen here at an Early Times picnic. *Photos courtesy of National Automobile Museum (below), and Street Rodder Magazine (bottom).*

Daddy". Strictly stupid! I never even went to see 'em but Henry decided to replace the "Big Ed" with "Big Daddy". I figured since I had five rug rats at home & since I was big (250 lbs at 6'2") the name fit pretty good. I guess over the years it worked good for me. The only other Big Daddy of note is Don Garlits the drag racer. Lotsa "Finksters" get the two of us mixed up! Anyway Revell was happy & so was I. Especially when the royalty checks came rollin' in.

The Outlaw was the first of a lot of car models to come from Revell. I sold it in 1970 to Jim Brucker for 50 bucks. He was a car collector & I figured if he got it he would make sure it got into the right places. I was right! The Outlaw was refurbished to it's original condition by Harrah's.

THE CRAZY PAINTER

PINSTRIPING - SCALLOPING - FLAMES

ROTH

*Custom PINSTRIPING

*Plastic LAMINATES

*Car NAMES

LUDLOW 15381

4616 SLAUSON Maywood California

CALIFORNIA
FHB 572
Studios
CAR & BOAT Show
SPORTS ARENA
CAR & BOAT Show
SPORTS ARENA

Fritz Voight

My parents knew his parents, we went that far back. Of course, I'm older than him. I was born in 1924 and he was born in '32. In those days there wasn't such a thing as gangs, the bottom line is, when you were sixteen a guy was two years older he's in a completely different group, so I never really hung out with Roth.

He always was kinda artsy. When I say artsy I mean building weird shit. He just loved that weird shit and that was the opposite of how I felt, I didn't care what it looked like, I just wanted it to go.

I was neighbors to the Drake Family and then I worked for Louis Meyer when I was a kid—I was crazy about cars. They were way older than me and I used to stand in the background and listen for little tidbits of information.

On Saturdays, when they'd load the race car up on Broadway Street and Walnut Park we'd go and sit across the street like birds on a fence and watch them. They'd take it up to Ascot to race. We left Walnut Park when I was 8 or 9.

Roth knew I was in the garage business and I used to help him with his cars. He was a little bit of a mechanic—he never did claim he was a mechanic—his main line of work was design and ideas. He'd go crazy with the plaster of Paris, wire mesh, plastic and crap—they'd have that shit all over—everybody that walked in his shop turned white.

Doug (Dirty) did a lot of the grinding—I can still visualise him with clear spots where his eyes and nose were covered in dust from grinding plaster of Paris. Evidently, it's not toxic because everybody would be dead. Nobody ever wore a face mask.

I remember walking in there once, when he [Roth] had some kind of car that looked like it had some rats pulling it and I said "Jeez what a lot of work for a bunch of shit"—but that was what he was always known for.

He did T-shirts to make money, they all used to be hand done, that's where he made his money and when he got that money he'd fool around with the cars.

I'd build engines for him. He'd tell me sorta what he wanted and I'd do what he wanted but I don't know which engines went in which cars. I'd not go down there and help him put them in, he'd stick them in himself.

Left: Ed's favorite tow car was this Dick Cook-built '40 Chevy fitted with an Olds Tri-power. Above: The Outlaw's Caddy engine and those of many of Ed's cars were built by Fritz Voight. ***Photos courtesy of Petersen Publishing (left) and Ed (above).***

I remember I worked on some chrome carburetors for him and they were a pain in the ass because when you chrome die cast you copper plate the hell out of it but it never sticks good and it gets on the threads and gets punky. But Ed is Ed and that was what he wanted.

I also remember he had a blower and a carburetor inside of the blower and I used to think that was ridiculous but he did show stuff. He'd always be dreaming up new ideas.

I was in the speed end of it and Roth was never in that end of it. He was into being different and artsy. So I can't say what the hell I ever worked on for him that went for what. I used to do his regular stock work even. I think I bought one his first Oldsmobiles and gave it to my mum who was just learning to drive.

I had another guy who worked for me for years that was a real good mechanic who used to do a lot of his work, a guy named Dick Cook. He was another one of those guys. He started with Roth and Roth's ideas rubbed off on him. He built a Firebird with two engines.

Cook was a very good mechanic and very creative in cobbling things and he was a good welder. Roth was a very punk welder. When he used to make those trikes out of VW chassis, he used to take a torch and cut through that malleable iron and leave a lot of slag and he'd try to weld right over it. I'd say "you can't do that, you've got to get all that slag off." But he was always so impatient. But when Dick Cook pasted all that stuff together he knew what to do. He worked for Ed and he worked for me at the same time. He was very creative. ❒

Beatnik Bandit

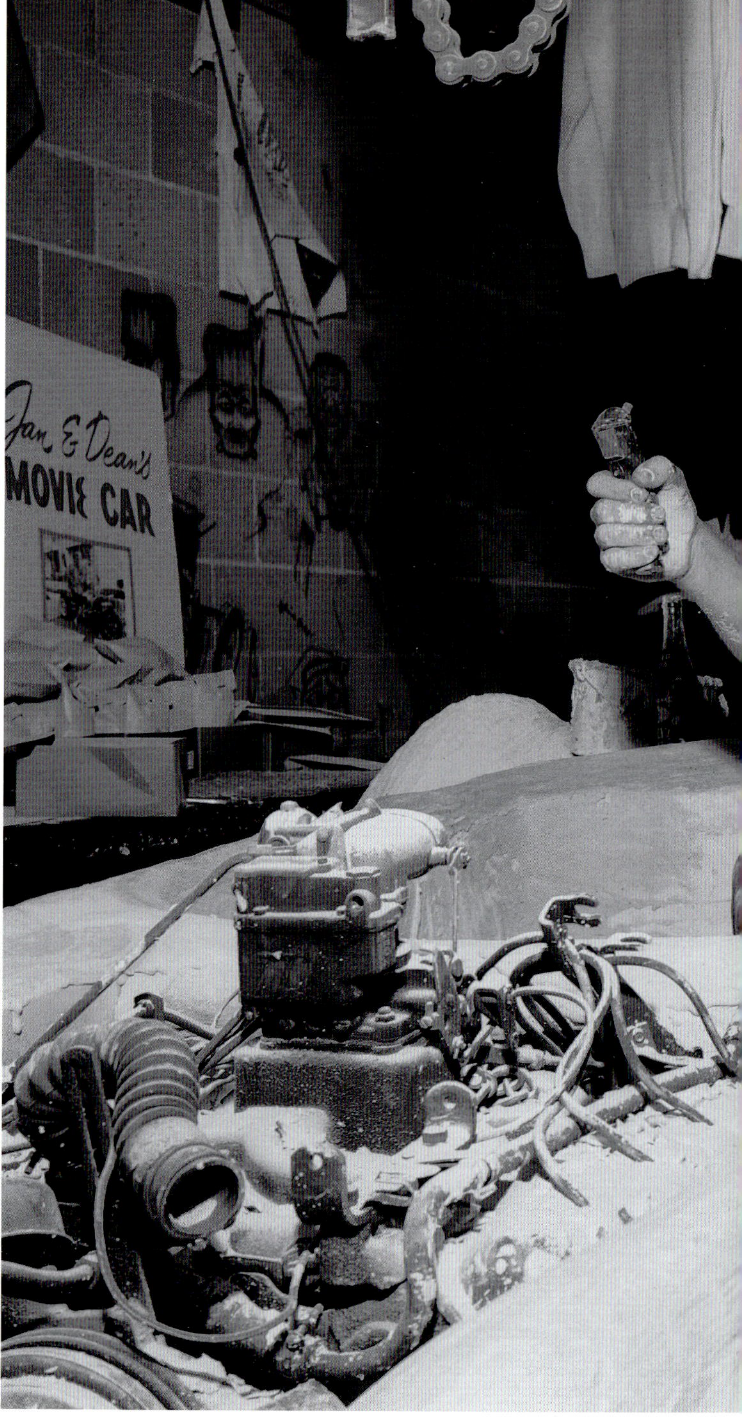

The Beatnik Bandit" started out as a project car for *Rod & Custom*. Joe Henning was one o' these crazy sorts like myself & he had this idea to build a ground-up future rod. So we planned a series of articles where we'd do some drawings & get some reaction from the readers. I needed a project & started building it. I soon found out the pictures Joe was drawin' & the actual facts didn't jive so we made a lotta changes during construction. Every time I'd make a change Joe'd make a different drawing.

That's when Doug Kinney walked into the shop. Doug had this tuxedo that he'd gotten married in & he'd wear it to work. He was always neat & clean so the name "Dirty Doug" stuck to him. Doug became the official sander & painter. He worked for me from about 1960 to 1969. Doug's this really shy guy that is a crackup whenever he decides to cut loose. Him & Ed Fuller kept all the employees in stitches. Ya gotta realize that in between the fiberglas & plaster I was crankin' out T-shirts like the "Rat Fink", "Wild Child" & "Mother's Worry". A lotta you baby boomers got shirts from my place with fiberglas dust on 'em I'm sure!

A little known fact about Dirty Doug is that when Bob Hirohata died his now famous Merc custom fell into unscrupulous hands & really got wasted. Well, Doug was this car nut that had his whole yard full of really stupid cars like, Honda 600's & '57 Cad Brougham's & Berkeleys & so he goes & buys the Hirohata Merc for like, 200 bucks. Remember, in them days, nobody really cared about wasted customs. So he salts it away in this garage & for many years just takes his close buds over to see it. It was in primer & had this giant dent in the hood like some demented dude had whopped it for spite with a sledge hammer. He finally sold it many years later when the garage owner threatened him with extinction.

Doug's yard was like a museum. His mom did not appreciate all the cars (The whole yard was full) so she reports him to the local fuzz! Get this! They comes down & make him clean up the yard! Can ya dig it!?!

By the time the Beatnik Bandit came along in '60 I had a set of wrenches & put a down payment on a "buzzbox" welder. Shop? Can ya dig the yard? Me & Dirty Doug worked under this corrugated roof shed (lucky that the sun shines most of the time in L.A.) that Dick Cook erected in the back of the shirt building. Then about the same time I erected this single car garage & we moved into that. It was tight quarters. Clean floor? Ha! Ha! Why clean it? It just got grossed out the next day. So we cleaned tha floor after we finished the car. I use'ta find screwdrivers & all sorts o' real neat junk in the

sweepings—especially screwdrivers. Lotsa fiberglas dust too. I loved to see them guys with the suits: FBI, Tom Wolfe, etc., come to this back room and freak out. Fiberglas dust is worse'n itching powder. Me & Doug wuz pretty use to it.

Most tourists that came to the Slauson Avenue shop were not hip to the strange way that me & Doug worked. It all seemed like one big plaster mess to outsiders, but we had method to our madness. Actually I don't know why me & DD are still alive. We've inhaled tons o' plaster & fiberglas fumes. DD use'ta take one of the Rat Fink T-shirts & slip it over his head, tie the sleeves behind his head & peep thru the neck hole. Underneath that he'd have this Mickey Mouse (sorry Mr. Disney) air breather. He'd look more like a UFO pilot than a leadsled jockey. He'd grind hour after hour on the bodies getting them shaped the way I wanted them. Every time we'd walk in the dust it'd be like this big puff of white powder. Yech! Yech!

In the back, under the Dick Cook awning, we had this pile of plaster & cardboard. People'd see it & back off real quick. My kids knew what it was. The dreaded "mistake pile". It was a collection of all the trials we'd make on the cars & didn't like. Y'see building the cars outta plaster was cheap & easy so me & DD would try insane stuff & if it didn't look right we'd take a carpenters saw, one o' my dads big buck ones, & cut off the offending piece & toss it onto the mistake pile. Then I'd take coat hangers (I used them for welding also) & split 'em in two & poke'em into the plaster & mix a new batch o' plaster & try somethin' different. That's how the insanity of the "Mysterion" nose piece happened. The mistake pile grew a lot in the old days. Then just before we'd paint the car we'd haul off the mistake pile to the dump to get rid of most of the dust. Then there was the scrap metal pile!

The scrap metal pile was just metal, but with all of the dust flyin' around it looked like it was plaster too. But if I kicked the metal pile with a good swift one I'd be able to see what kinda metal I had. It wasn't one o' those neat kinda piles. It had old lamps & electrical conduit & all kinds o' goodies that I could use for makin' new stuff. Sittin' here in this clean room

The beginnings of the Beatnik Bandit sketched by Joe Henning at Ed's request. Ed's plaster interpretation saw the T-top give way to a pizza oven-formed bubble top but apart from that remained pretty true to the original concept. *Photo courtesy of Petersen Publishing. Artwork courtesy of Joe Henning.*

at this word processor & thinkin' about all those years I'm havin' a spazz attack like I use'ta whenever I wuz dealin' with those two grossed out piles. I've got the same two piles goin' nowadays but I keep 'em clean. My plaster piles are almost non existent 'cause I work out the designs in my head first. Saves a lotta hackin' up o' those big lungers. My steel piles are mostly new metal. That's one o' the advantages o' bein' a retired (gimme a break!) genius.

The Bandit was a 1955 Olds chassis & engine that I chopped & slammed. The body was built with my plaster & fiberglas method. The idea for the bubble top came from Bobby Darrin's "Mac the Knife" dream car. I called Detroit to find out what a bubble top'd cost & they'd tell me thousands o' bucks so that I got this idea from Louie Aguirre to put some regular plastic in a pizza oven & then blow it up like a balloon while it was still hot. It was a hot idea (Get it?) It never failed. First time, every time. Plastic was still in it's baby stages & the plastic'd break real easy so we really hadda be careful to bolt it to the lifting frame, I used rubber sink washers.

I was always real hip to what wuz comin' down at the drag strips. I use ta go & set up my T-shirt booth on Saturday nights at Lions in Long Beach. I made my kids stay at the booth while I made my way through the pit area. I watched what guys wuz usin' to make their rides go faster. Blowers & multi carb manifolds, usually hopped-up 97's, and slick racing tires were the hot setup on race cars & dragsters so I used the same basic stuff on the Bandit. I had sent it all to the chrome plater & when I put it all together I couldn't get the sucker to crank over & ignite. So I calls this same Fritz Voigt & he brings his screwdrivers & wrenches & in like, 10 minutes he had that mill eatin' right outa his hands. There was plaster dust & exhaust smoke all over that garage. We all scattered like rats! Dirty Doug went outside hackin' his lungs out & Fritz split back to his shop mumblin' somethin' about us car customizin' guys that wasn't nice at all.

My car show commitments got bigger & bigger. And Revell was hittin' me up for more cars to make models of. The east coast car show promoters was callin' me beggin' for new show cars for their shows. The Bandit was the first non practical show car I built. I had this one center stick that controlled the steering, speed, & shifting of the whole car. It also became the first show car that I hauled on a trailer 100 percent of the time. It ran, but it was a tricky handler.

Paintin' cars wasn't my favorite trick. If ya don't stay on top of the latest technology the results can be disastrous. Well in the old days,'59 or so, I painted my own stuff. It was simple. I painted the Outlaw white & then got some clear & put in a spoonful of this yechy white pearl essence stuff we got from the local taxidermist that came from boiled fishs cales. Mixed into the clear lacquer it looked like a loogey highball. Squirted on top of the white it became known as a "Pearl Job". (Taxidermists used it on mounted fish.) Then I'd put the scallops on top of that. In the case of the Outlaw I used a combination of green & blue toners & painted the scallops over a silver base. I have had people tell me that the covers of *Car Craft* make the colors look like they're purple. Optical illusion I'm sure! But after the Beatnik Bandit I knew that I couldn't stay up with what was goin' on so I got hold o' Larry Watson.

Watson had this shop on Lakewood Boulevard & he started out with me & Dutch & Jeffries pinstriping cars. Watson had this really tough '49 Chevy that he'd painted

Left: According to Ed, the tape measure was a publicity stunt, all the Petersen guys knew he never measured anything so photographer Eric Rickman asked him to pose with one. Ed's original idea was to cut doors in the body but he found the glass difficult to work and settled for a wooden ring lifted by convertible top cylinders and pump. Below: From left to right: Ed Fuller, Neil Brown and Richard "Asheye" Ash knock out plaster. *Photos courtesy of Petersen (left) and Griff Borgeson (below).*

with this outrageous candy purple over silver & whenever ya'd scope it out in the sun it'd almost knock out your eyeballs 'cause it was so bright. The effect came from painting this mixture of clear lacquer & purple toner on top of a silver base. About 20 coats of paint were necessary. The end result was awesome.

Watson had a sales gimmick that he thought up to sell those paint jobs o' his. He had this four foot by four foot plywood board with about a hunnert light bulbs screwed into these porcelain bases. Each light bulb was painted a different candy color. Wow! Customers couldn't refuse one o' those high tech, high dollar jobs after seeing those light bulbs. I was the same way. My problem was I didn't have the kinda bucks Watson was chargin' so I made a deal with him that he could take all the time he needed, usually a month, & I'd trade him Rat Fink T-Shirts for his work. He went for it! I always thought his best job was on the Mysterion where he painted candy yellow over a silver base. I always tried to give him credit on my signs at the shows & now through this record it remains a fact.

I hauled the Bandit with a Cad hearse. I ate & slept in that hearse. It's been to a lotta cities at a lotta car shows in the '60's. By 1970 the Bandit had been painted

More Bud Lang shots from the Bandit's debut. The frame was a '50 Olds shortened to 85 inches, the Fritz Voight-built Olds engine featured a Bell Auto Parts blower with twin Ford carbs while the fender-mounted antenna operated the bubble top. The interior featured joy-stick controls and Ed Martinez-upholstered seats. *Photos courtesy of Darrell Zipp (color) and Joe Henning (B&W).*

& changed a lot on account o' the many dents & scratches it got travelin' in trailers & bein' shoved around at the shows & so I sold it to Jim Brucker for 50 bucks. He then traded it to Harrah's in Reno in about 1973. Harrah's restored it in 1985 to it's original shape. Today (1994) it's on permanent display at the National Automobile Museum in Reno, Nevada.

Originally I wuz naming my cars after western cowboys, so's "Bandit" was the original name. Then I went to Kansas City for a show & while I wuz cruisin' thru town, tryin' to catch me some grub, I picks up on this newspaper that's tellin' about this bearded dude that robbed a convenience store so the reporter called him the "Beatnik Bandit". I picked up on it & changed the name.

Naming my cars has always been a mind boggling experience to say the least. I realized a long time ago that a good car name is just like a good nickname for a pinstriper. When us guys wuz buildin' rods at Bell High

School we never gave 'em names. I think it was Norm Grabowski who started that bit with the show cars. He had this real radical T roadster & there was this TV program named "77 Sunset Strip" that was photographed at Dino's (Dean Martin's) nightclub on Sunset Boulevard. Turns out that the parking attendant was a dude named "Kookie" who was a real savvy, hair combin' hipster & the producers of the program needed a car for Kookie to drive around in. Norm's T was perfect. So Norm went to the car shows with "Kookie's Car".

By the time 1958 rolled around almost everyone was puttin' names on their cars. Me too. Sometimes I wonder if the "Beatnik Bandit" would be as popular as it is if I'd named it "Litmus Paper" or sumpthin' like that. I'm lucky! I know Robert Williams. Robert worked for me like, four years in the 60's. He became famous as an underground artist later on but at that time I found out that he had this uncanny way of thinkin' up stuff. He's named "Captain Pepi's Motorcycle & Zeppelin Repair," the "Conastoga Star" & "Druid Princess" & helped with other stuff I've had problems with.

Working the shows, we all got to know each other. We did a lotta hollerin' back & forth & with Mouse, if he was really far away, I'd get kids to take notes like, "You stink" & he'd send back stuff like, "I hate rats" or, "You're armpits smell." The kids transporting these notes got a kick outta this, & it added life to the show.

Some years later Mattel calls me & wants to make this new series called "Hot Wheels" & they wanna start it with the Beatnik Bandit. We agree on a royalty fee (I think it added up to about 800 bucks) & it was very very good to them. In 1992 the head honcho calls me & axes me if they can re issue it to which I replies, "Not interested," 'cause they wuz gonna repeat the original mistake and make 'em in red & blue, etc., when the original car was pearl & candy red.

Next thing I know guys're bringin' 'em to me at car shows to autograph. No way man! I wasn't gonna sign any unauthorized models & some o' my fans really got excited & angry at me. All's I can say is, "Call Mattel & complain to them." Maybe they'll tell ya the same thing they told me, "Well, what'cha gonna do about it?" So do me a favor & don't bring those unauthorized models to my booth.They wanted to settle outta court by givin' me 500 of the models. Who wuz they kiddin'?

Robert Williams

I had been fired from the Weyerhaeuser Corporation, down in Maywood, in '65, where I was a container designer. It was desperate job—and I went back to the unemployment agency like I always ended up going and they said, "We got this one job that everyone we've sent down there has rejected."

"Well, what is it?" I says. And the lady says, "Its being the advertising and art director for Ed 'Big Daddy' Roth down in Maywood."

"Hey," I said, "I was made for this job. Give me the phone." So we called up Ed and Ed said, "Come down and show me your portfolio."

Ed looked through my portfolio and says, "If I'd known you existed, I would have hunted you up."

I had first met Roth in 1960 at the Civic Auditorium in Alberquerque, New Mexico. I was at a car show there and the Thursday before the opening me and my buddy come wandering in the back door and the Beatnik Bandit was sitting right there at the entrance and it just knocked me out.

It had cordons around it. Roth was nowhere but there was a bunch of employees there and the promoter of the show, a guy named Gene Bender. And we're just two kids, me and my buddy drooling over this thing. I had my foot on one of the stanchions and Bender says, "I don't think Mr. Roth would appreciate your foot on that stanchion."

They all walked away and we're just standing there and my friend crosses the cordons and he spits a big lugey right down one of those open throats. And I just flipped.

He had just finished the Wishbone when I got there and he'd started on the Druid Princess. He didn't like the Wishbone from the very beginning. I didn't like that either but Ed had to turn out a car maybe every eight months and to do that he had to have such a compromise factor. He started out making fiberglass bodies for the Outlaw with male and female molds, but for his second car he started putting it around plaster and not knocking the plaster out. There was no male or female so you know those people that restored those cars at Harrah's got inside that fiberglass and realized, hey, this is the matrix and everything of this car, there ain't no female.

He found a plaster that had a fiber base, it was kinda of a fluffy plaster, so if you left big hunks of it inside the body it didn't matter. Those pods that were on the side of the Road Agent, they were solid, those were never cut open and the plaster taken out.

Ed was tied in with Revell and he let them talk him into a contest. So Revell selected a winner and it was this kid who had all the specifications: It had to be this kind of a frame and it had to be a Buick engine in the rear and Roth was obligated to actually build this car.

I didn't know about this and I was in the back one day stumblin' about and I came across this weird chassis with this Buick engine in the back and it just looked like the silliest thing and I asked Fuller, "What's that thing?" And I got the whole story so I dug around and I found the original model car that the kid built.

Anyway, Roth got tired of the project, he just had better things to do and the kid's family raised hell. And I remember a letter came from the kid's mother and it said, "My son's now in Vietnam, he's a grown man and you haven't built his car. You'll never build his car. We've reconciled ourselves to that but would you, out of common decency, please return his model."

There was a testimony there. I loved Roth. I loved that whole world but there was a flakiness there also. That said, he's also the most giving person. I've seen him give away just incredible things. He was just a remarkable person in that respect.

Roth used to have his extra clear bubble tops and the Outlaw molds on the roof and I says, "Ed, don't ever throw those Outlaw molds away. If you're gonna do something, give 'em to me, I'll take care of 'em and give 'em back to you if you ever need 'em back. And he hemmed and hawed and then one day Jake came up to me and he says, "Listen, Roth gimme these body molds, I don't want 'em, would you come and get 'em?" I've still got 'em.

Jake came in with Dan Woods as kinda part of the bargain and when he and Jake were there that place was a real hot rod hub. Right when work was over everyone would start comin' around cuz it was just a party. A lot of the people quit coming around when Roth got into the biker stuff. It didn't seem like their world much any more. But then the motorcycle speed equipment people started coming in because Roth had the only book that catered to the biker thing. Until then they had nowhere to advertise. So he started a whole industry, which people don't realize.

He got into the bike thing the year after I had started to work for him. He kinda graduated into it. One thing you have to realize, Roth gets in these social deals and he doesn't look at it like we do because Roth is six foot six, weighs 250 lbs looking thin and he's a black belt. Roth does not have a fear quotient like we do. When Roth talks to somebody he talks to them out of pure honesty, he doesn't worry about what they're gonna think or anything. ❐

Left: They came in their thousands, here to Montreal, Canada. Demonstrations on the hour, every hour. Top left: Billboard in Buffalo, N.Y., 1962. Right: The show circuit was equally tough on the Bandit so Ed repainted it green. ***Photos courtesy of Doug Pizac/Bob Hegge Collection (left) and Mike Lowe (right).***

Rotar

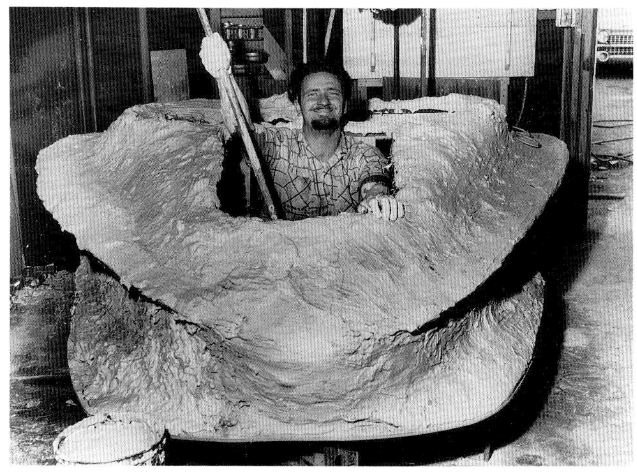

The biggest battle of all at the car shows was with George Barris. It was too serious to be funny. I mean he was the head man. Numero uno. The guy way out front. We had both built air cars & at first we'd just set 'em in the shows & not demonstrate 'em but the heat from the public put us both on the spot. Whenever I'd unload "Rotar" (Roth Air Car) at the shows & the newspaper reporters were there I'd make a big point of telling them that I'd demo Rotar & it would lift higher than the Barris air car. It got hot at times 'cause Barris & I were miles apart in our thinkin' in those years. It was a unfriendly feud that stopped when Rotar blew up & scattered parts into the crowd as Ron Aguirre was doing the demo at Cobo Hall in Detroit in '64. Barris won! My eternal apologies to those people, especially an unnamed lady who suffered for many years after.

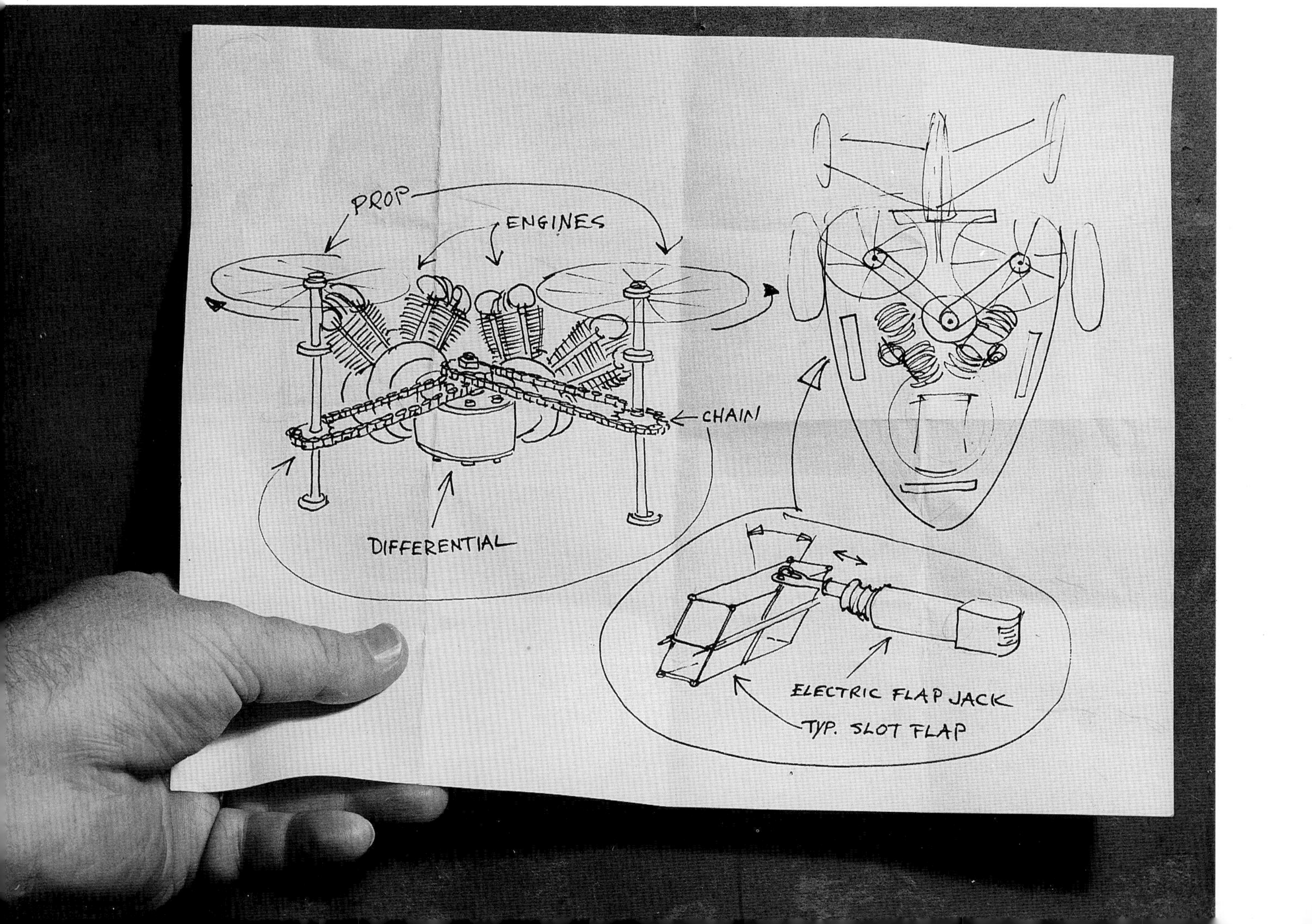

George's air car was a well built pretty little thing but Rotar was on the rough side with two Triumph engines turned on their sides, each with a high-pressure

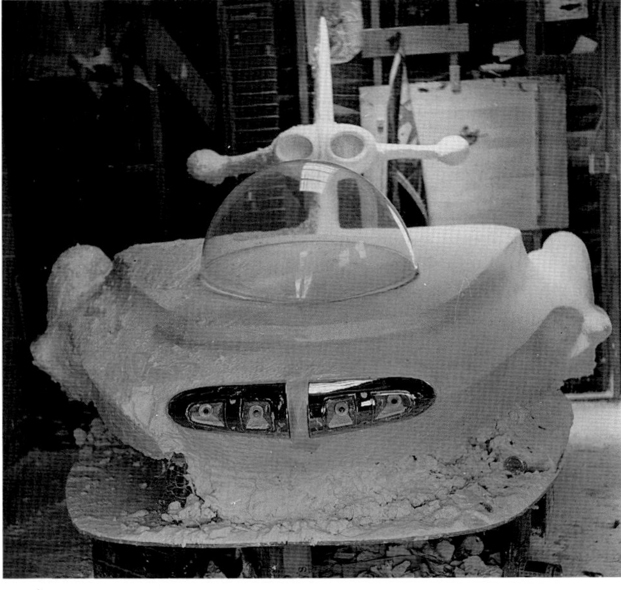

prop that created a cushion of air that allowed Rotar to float on land as well as water. My directional flaps on the bottom were plywood (unpainted yet!) and the rear fin was fashioned after my '59 Cad. Since the top bearings of the motors weren't oiled, I forgot the dry-sump trick, the bearings froze causing the crank to snap and the prop to take off into the crowd causing a bloody mess. Rotar is currently owned and being restored by George Goodrich.

The stories about me dealin' with sleepin' in the front seat of my car & havin' the flies buzzin' around my stinky feet & all the other rumors about me smellin' up the show with armpit odors are all true. I never had bucks for things like motel rooms. After all the 20 bucks ya drop on a room can get ya another carburetor or pay for some more chrome. It is a sickness, but I'm happy!

Left: Rotar's drivetrain concept depicted V-twin engines but Ed eventually used two Bell Auto Parts-built 650cc Triumph twins laid on their side. ***Photos courtesy of Ed (left), Petersen Publishing, and Fred Caldwell (Cincinnati show poster).***

Left: Rotar's tailfin was fashioned after Ed's '59 Caddy. Below: The smile on Ed's face belies the problems to follow when one of the engines snapped a crank causing the fan to injure some spectators. ***Photos courtesy of Petersen (left) and Ed (below).***

Which reminds me of a very funny story. I wuz cruisin' down in Louisiana with the hearse (originally I had a '57 Cad & then I bought a '60) & decides to pick-up on some gas. I pulls into the station & this superstitious dude starts tellin' me he ain't gonna serve me & axes me to leave. I couldn't figure out why he was so nervous. So I gets out to check my rig & discovers that the back door of the hearse had some of the small foam rubber pieces that I used as helmet liners for the German "Surfer" helmets we sold at the booth & in the dark they looked like fingers of a dead hand. No wonder the guy was nervous!

George Barris

During that fifties period Ed started coming in. He was doing striping for me and he did shirts for me and he wanted to finish up the Outlaw at my place over in Lynwood.

He had his own feeling of what he wanted. Ed is a creator, an artist and the big thing about Ed is that Ed was not a customizer for the public, he did work on cars for himself. What he did for the public was sweatshirts and T-shirts. The cars he built mainly for his own creativity, for exhibition and to create product awareness for himself to get his name going, so the competition was not customizer against customizer. He was creative in his own world whereas I was doing styling for the public world and the film world so there was no real competition and that's why we were comfortable in our rivalry—it was all part of the show. He would show up with his wild stuff and I'd show up with my stuff and the public loved it.

The only thing we had a real battle over was the air cars. I built the X-Pac 400, which was a ground-effects car—you could fill up the plenum chamber with air and then it would come up and float.

Trying to get it to go in different directions was the hardest thing because nobody knew how to make them movable. The way I made mine move in 1960 was I had little thrust blowers on each side and mine was all remote control. Then Ed built his Rotar and said, "My air car will do this," and I kept saying, "Mine will do this."

The battle of the air cars was one of the great show car spectacles of the early 1960s. Above: George Barris proudly displays his air car the X-PAC 100. Below: Rotar as it appeared in 1962. *Photos courtesy of George Barris (above) and Bob Lewinski (below).*

Because we were doing exhibitions I had to use electric motors and batteries and the weight was really too much for the electric motors available at the time. We altered the pitch of the fans, then I put a skirt on the bottom so it did lift up to the full length of the four-inch skirt then I put a parachute underneath so it would blow and blow girls' skirts up.

That was the big deal—the battle of the air cars and it was very good, the public got a big kick out of it.

Sadly, I had so many cars here at that period I had a guy store the X-Pac for me out at the airport and the guy died and I never knew what happened to it. ❐

Old Pro

I bought "Old Pro" as was in 1962 when I was bucks up from all the royalties Revell was sendin' my way. I'd always had this fantasy about ownin' a Model A truck with a Chevy engine in it & so I told my scouts to be on the lookout for a cheap A pickup. I wanted to get a stocker 'n do the hot roddin' myself. I figured it'd run about two grand to get the job done.

I kept gettin' reports about this cute '29 for sale—cheap. The guy was tired of it 'cause it seemed to ride & steer badly. It had all the Chevy muscle under the hood & the dropped axle & hydraulic brakes and big & little skins just like I was gonna install. I knew the steering problems came from a lack o' some decent shocks in front. It was exactly what I wanted and the price was right. Otherwise I never woulda went for it. I put on the shocks and it boogied right along.

I drove it for six months or so & installed a remote-control starter on it. In other words, I went down to the model airplane shop & picked up a six-channel radio control transmitter (high tech in them days) & made it honk & start (headlights too) from a long distance. Me & Doug'd park it in the parking lot at the shows & freak out the citizens in the place.

Later I sold it to some poor unsuspecting dude (only kiddin') in Salt Lake City who, I believe, still owns it.

Bob Larivee, Sr.

I got to Ed and made a deal with him in about 1961 to bring the Outlaw back to a bunch of shows. And of course the stories of those trips of Ed's are really something—I wish I kept a diary. Remember, I was a very conservative guy at the time with no exposure to a person like Ed. He was the wildest free thinker I could ever have imagined. Anyway, Ed brought the Outlaw back and it went over real well and he did our whole series of shows.

While he was on that circuit I made a deal with Ed. I told him that at the end of his tour that year I would like to buy the Outlaw. Ed agreed and I think I paid him $3,250 and he was real sad about selling—I guess he never thought I'd come through—but he saw it as an opportunity to simply go on and build something new because the cars really had little meaning to Ed aside from giving him a chance to show his creativity and earn him the money to do other things. They were really the vehicles with which to promote his T-shirts.

Ed had already started on his new car, the Beatnik Bandit, and when that was finished he already wanted to built the Mysterion, so he sold me the Bandit. He didn't tour the Bandit like he had the Outlaw. The shirt thing was really taking off. He had a million things going—that little shop on Slauson was a mad houses. So we took the Bandit out.

Then I got to thinking the way he did: Why did I need these cars because now we were looking at them just as feature attractions and I love those cars. I mean, I knew everything about them and of course the Bandit was really unique, it was a bubble top, remote-control car so we could do these demonstrations where we could take the top up and down and the public had never seen anything like it.

Both cars were big hits but Ed already had me excited about the twin-engined Mysterion. After a lot of soul searching—the value of the cars was going up—I decided to trade the Outlaw and the Bandit back for the Mysterion, which was probably the worst deal I ever made.

I was happy with the deal at the time. I had actually got my use out of them. Ed was never an engineer and structurally his cars were not very built from the point of putting them in a semi-trailer, and this was before electronic air-ride suspension, so the cars rode rough. With the Outlaw and Bandit we didn't have much trouble but as we got into bigger trucks and loads, the Mysterion was a real problem. The frame was extremely wide to accomodate two engines and there wasn't enough engineering in the frame to keep the thing together. Welds were giving and the thing was forever breaking apart. One week we'd fix this part, then next we'd fix that part. We were forever repairing the thing. I think we even took the engines out and welded it up and reinforced it and chrome plated it but it was basically not a well-designed frame, so when we were done with it we didn't know what to do with it. So I called Ed and told him how bad it was. He was ready to let us have another car but the Mysterion was in such bad shape we basically junked it. And that is the last I saw of the car. We took the engines out and hauled them back out to California. I don't remember that we gave him back the body. It might be that we did but I know we scrapped the frame. You just couldn't do anything with it, it just wouldn't stay together.

In the meantime, he'd finished the Road Agent, which was a whole different kind of car with a real lightweight frame—I think he made it out of exhaust pipe—but he had no weight in it. It was rear-engine VW, I think, and so we hauled that for a couple of years and we had some structural problems but nothing like the Mysterion. Eventually we gave him the Road Agent back and I ended up without any Roth cars.

At about that time the last car we did a big number with Ed was the Rotar, which became notorious here in Detroit. It was the battle of the air cars. Barris had built the X-Pac. The difference was Ed wanted this concept to work and so he was crankin' those things and he'd actually get the car to move a little bit when the crankshaft broke and this blade took off and part of it went off into the audience. I can't imagine what would happen today if something like that were to happen— a couple of people were hospitalized and Ed went to visit them in hospital. Our insurance covered everything and it wasn't any big deal but that was 30 years ago. Not that there should be anything good about a disaster, but we made the front pages of every paper and we got a tremendous amount of publicity, and even though the car never ran after that, Ed hauled it around and so did we.

I know he built other cars but when he built the Druid Princess we parted ways with Ed. He had a million others things goin' on, the Revell thing, the biker thing and I guess he didn't really see the shows as having a lot of longevity. At that point he'd been around the show thing for ten years. ❐

Tweedy Pie

As the royalties from the Outlaw model started rollin' in I found myself able to get all kindsa tools & fiberglas & junk parts for a new car. I even got new furniture & stuff for the pad but my free time was gettin' less & less. I was spending more time at meetings & car shows & less time at home or working in the garage.

When I was goin' to car shows with Little Jewel and the Outlaw I started hangin' out with Bob Johnston who had this cute T-bucket named "Tweedy Pie." Bob was from the old camp of hot rodders who drove their rods to school and kept 'em clean. Bob wasn't into trophies or publicity. He just enjoyed hot rods.

Tweedy Pie had a flathead Ford engine which was plenty cool until 1955 when Chevy came out with those Duntov-designed overhead mills. Bob just kept his ol' flathead (it was 1962 yer know) but he knew in order to stay on top of things he'd soon hav'ta get an overhead engine.

Bob was an instrument mechanic at one of the local airplane factories & Tweedy Pie was outfitted with the best gauges known to man. He'd get a blast outa explainin' the inner workin's of all those fancy dials to me at the shows. He wasn't really into cars & after he sold Tweedy Pie to me I don't ever remember seein' him with a rod.

Top buck for a rod in them days wuz about $1,500. I made him a deal he couldn't refuse. I finished the Chevy installation & put dual headlights on it & took it to the shows. Revell made a model of it. I sold it to some dude in Sierra Made, Calif. a couple years later and I think he still owns it. He put some "Mickey Mouse" New York fenders on the rear & I hardly ever see it at the shows anymore. I learned a big lesson about sellin' to individuals—If ya wanna car to get changed (yech) & hidden in a garage—sell to a private dude.

Top: Originally built by aircraft instrument mechanic Bob Johnston, Tweedy Pie was a 1920 Ford T-bucket widened 3 inches and channeled over bobbed Deuce rails and powered by a flathead as shown on the cover of the December 1959 cover of *Rodding and Re-styling. Photo courtesy of Jim Ballard.*

Before selling to Ed, Bob swapped the flattie for a '57 Corvette motor. Ed, who had striped the car for Bob, subsequently added an Offy manifold and six carbs, homemade nerf bars and lotsa chrome. And don't you love those shoes? ***Photos courtesy of Ed.***

IGNITORS
NKE 687

Mysterion

I got the idea for the "Mysterion" from the dragsters. The dragstrip was seein' a lotta dragsters poppin' up with two or three or even four engines in 'em. It was all started by this really neat Hollywood child star "TV" Tommy Ivo. Tommy had this talent for shakin' the ground (& people) with his craziness. I figured I'd build me a show car with two engines in 'it. Trouble was I didn't have bucks for twin engines & drive train. I was runnin' around at the shows with this harebrained dude named Bud "The Cat" Anderson. Bud was in charge of the Ford "Custom Car Caravan". He'd make sure all of Ford's latest stuff was in the car shows around the country. Y'know, white shirt & tie & all those downtown big words to impress the Ford bigwigs. I razzed the snot outta poor Bud. I'd scream & holler at the shows like he was my mortal enemy & he'd always have this sly answer. But after hours we'd go to dinner & laugh about the whole scene. Bud got me three brand new Ford mills with trannys.

Three! In them days the factories were givin' away stuff like there's no tomorrow. In the back o' my mind I needed a new tow car. I was gettin' a lotta flack about the '40 Chev & figured I was ready for a new tow rig. Ak Miller was the local Ford tech advisor since he won a recent Baja race with a Ford powered somethin' or other. Ak also put in the good word for me. I had no intentions of usin' a Chevy tow car for the third engine. but I gotta tell ya as great as them Chev mills were, the trannys were junk. I hadda have a good tranny to tow the trailers with so that's why I ended up installing it in my '55 Chevy beater. It was lighter than the Fords & lookin' a whole lot better. The other two Ford engines were used in the Mysterion.

I made this custom rear end out of an early Ford differential. I welded two rear ends together & welded the spider gears on one of the ring & pinions solid so's I could operate one or both engines at a time. I just figured I'd do that instead of doing the Tommy Ivo trick which was to mesh the starter rings with only one drive shaft (I had two of course). I actually started out with the frame & then I made the front of the grille with the offset lights.That turned me on so much that I was able to finish the rest real easy. Larry Watson painted it candy yellow.

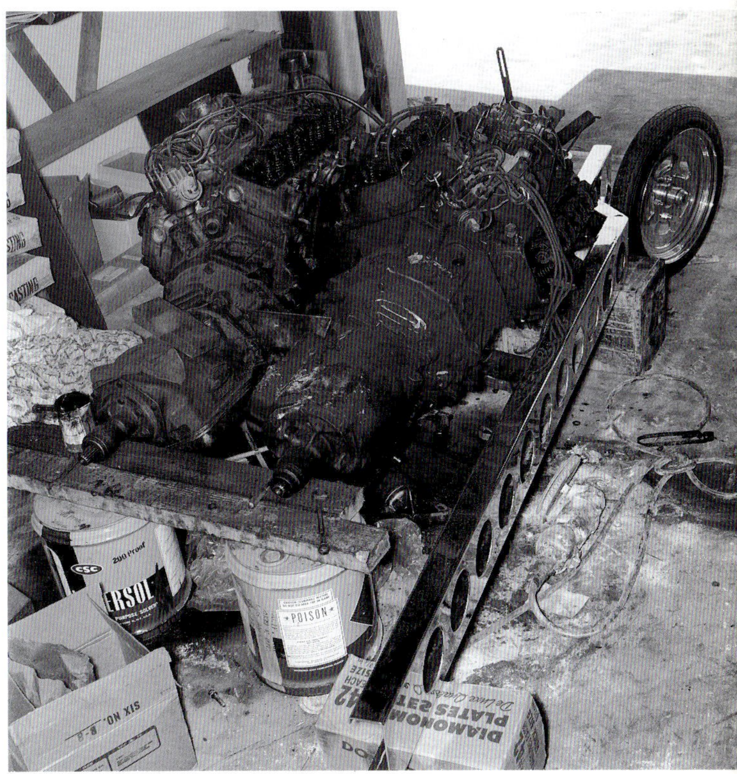

Left: Tweedy Pie at the Revell plant being measured for model making. Note the twin headlights. Inset: Ed sold Tweedy Pie to Chris Lovoy who made his own modifications. Right: The Mysterion takes shape perched on cardboard boxes and cans marked poison. *Photos courtesy Ed (contacts), Street Rodder (inset), and Petersen (right).*

I took it to a lotta car shows & then rented it to car show promoter Bob Larivee & while he was haulin' it around to car shows he noticed the frame starting to crack. He took the engines apart & removed the crankshafts & pistons 'cause he figured that was the problem but the problem was that we had chrome plated the frame without stress relieving it. It was in them years when chrome was so new no one was worried about hydrogen embrittlement. Nowadays, all ya do is put the chrome parts in an oven at 400 degrees for two hours right after it comes outta the chrome tank. No more crackin'.

So Bob calls me & tells me the frame is zapped & what should he do. I says "Pay me for the damage & I'll pick it up later." But later never came & so it ends up in Tulsa in some guys garage, where I will eventually have to reclaim it as soon as the cops find it.

Ya gotta remember in them days there was no sucha thing as goin' down to your local Super Shops & pickin' up on tires or rims. We hadda get the slicks from this place called Inglewood Tire Co. They made tires for dragsters & what they'd do is take a stock wide whitewall & put this humungous band of rubber around it & with heat apply it to this stock tire. It was pretty grossed out but that's all there was. As far as rims was concerned, we hadda make our own. Now I had done this for the previous cars. I was usin' Harley Davidson motorcy-

Left: Ed says the Mysterion nose was finished way before the body and the front axle was filled with solid weld! Below: Twin drive shafts dictated only one seat while a tape played during shows such as seen here at the Mid-America Auto Spectacular, Topeka, Kansas, about 1964. *Photos courtesy of Petersen (B&W) and Roger Kilborn (color).*

cle tires which happened to be of a 16 inch variety. Trouble was that they were too narrow for the car rims & the bike rims had spoke holes in 'em so I narrowed 16 inch car rims. When I made the Mysterion front wheels I wanted them super narrow. I also made & hand filed, not with a grinder but filed by hand, the spiders. Then I welded it together with the rim & had the whole wheel chromed. When I got the wheels back from the chrome shop I almost never got the tires on 'cause I'd made the rims almost too narrow.

A word about the guys at Model Plating. Y'know when normal people bring them stuff to be chrome plated the guys polishing usually know which side of the stuff to polish & buff. Not my stuff. This lady (Irene) she use ta cringe whenever I came thru the door carryin' all the outlandish brackets & rims. Finally they hadda draw each part & which side showed on the car. The guy at the polishing wheel can make or break the chrome job. This big Russian dude (named John Dootoff) loved (?) my stuff. But Luke (head chrome honcho) knew the publicity was worth the extra effort. Luke's kid Richard, who use'ta suck on his baby bottle when I first started goin' there, runs the shop now. It's still at the same place in Bell Gardens.

Later on I used this harebrained guy named Mando. He'd trade me chroming for a flame job or signs or whatever. He had a more basic approach to business. Instead of big expensive machines he'd make his stuff outta old car and washing machine parts. He'd pull in some of his pals from Tijuana & use old sheds for his shop. But he did good cheap work.

Knowing that the Mysterion was lost and gone forever, designer Thom Taylor came up with this concept for a Chevy LT5-powered Mysterion II. So far nobody has stepped up. *Artwork courtesy of Thom.*

Perhaps the mystery of the Mysterion was finally resolved with a call from Doug Wright—the last known owner (see right). Here it's looking fresh at the Revell plant belying the problems to follow. *Photo courtesy of Ed.*

Doug Wright

My part of this history was very sad. It occurred in 1967 when I noticed what was left of the Mysterion resting alongside a wall inside Ray Farhner's shop in Rayton, Missouri. Ray informed me that Promotions, Inc., had finished touring the car and, knowing it was broken and in a mess, Roth was no longer interested in it.

Ed had apparently asked Bob Larivee to remove both front and rear axles and wheels and ship them back to Maywood. He didn't care what happened to the rest. When I asked Ray what he was going to do with the car, he said, "Part it out."

My senses went crazy. Lying before me half butchered was one of the wildest, most famous show cars ever. The small TV was missing along with some of the rear hydraulic system. One of the Ford C-6 transmissions and one of the small radiators were also gone but other than that the car was complete, so I made a deal with Ray and bought it for $750.

There was no title but Roth required me to sign a legal contract stating that I could not use the name Mysterion or the name of Ed Roth in connection with the car. The photo you see here is the last picture take of the Mysterion.

Both engines had been completely gutted and their internals were stored in a box. I was told this had been done because the excessive weight of the engines had been too much for the upper front coil spring supports, which had broken while the car was being trucked around the country. When I checked the engine numbers, I was surprised to find that they were 390s and not the 406s as advertised. Oh well, that's show biz!

I was very happy with my purchase. There was a driver's control panel as well as a small outside control panel used to handle all of the hydraulic and other functions of the car. These included opening and closing the bubble top, and the hydraulic height adjustment of the rear suspension. Even the lid of the panel was electrically operated by radio control.

Close examination of the body revealed Ed's method of construction: the fiberglass was 3/4 of an inch thick in some places and the paint, which was now lime gold metalflake, was 1/16 inch thick.

I would spend hours sitting inside the car with the bubble top closed dreaming of actually driving the Mysterion down the road. At first I believed that I could restore it but I was not in a financial position to do so and after three years of recurring demands from my parents to, "Get that piece of junk out of the basement," I ended up having to do the one thing I had wanted to prevent—parting it out. Nobody wanted to buy it. The last to sell was the body and frame to a man who owned a body shop in Independence, Missouri. And that's the last I saw of it. ❒

Road Agent

The "Road Agent" started as a project car for *Rod & Custom* Magazine because R&C was tryin' to pump up its readers to be more aggressive and modern. T-buckets were "out" & more guys wuz buildin' customs with overhead mills in 'em. At the head of this project was Joe Henning who I'd worked with on the Bandit project. Joe had drawn this real slick "Rod & Custom Dream Car" in the mags but it had a front engine in it. I explained to Joe how front engines were out & Corvair engines were in. Joe freaked out 'cause Indy cars were his thing & his whole face turned red.

I went out an got me a junk 'Vair & built a frame out of 1-3/4-inch chrome moly 4130 tubing. I used this high-tech tubing 'cause they were startin' to use in on dragsters 'cause it was so super strong. I mounted the engine in the center & for a front axle I used a Bell tube axle with a modified coil spring suspension. I got the frame together & the thing ran backwards so me & Dick Cook turned the automatic tranny upside down and it ran like a charm—you shoulda seen all the tubes and plugs for that trick.

I called Joe back and he consented to do more drawings. I had always like rear-mounted engines 'cause it allows a lower hood line which makes for increased visibility. Later, Nader said they was unsafe. Phuuuut! On that noise!

When the body was layed up in plaster & covered with fiberglass Doug suggested we paint it the new candy apple red which had just been invented by Joe Bailon in 'Frisco. Dirt had painted all the cars to that point but he didn't know from candy apple so we paid Larry Watson a visit. watson knew candy & it seems to me he painted every car after that.

We did the "Road Agent" bubble the same way—pizza oven at Furst's sign shop—but used a fluorescent plastic which it still has today. I built the entire Road Agent without figurin' a place for the gas tank. Seems we put a surplus army tank in the deck lid at the last moment. The gear shift handle was a ratchet wrench. It had class!

I mighta mentioned before that I traded going to car shows for a booth to paint my monster shirts. I covered most o' that in my other book. I wanna stress here that the show promoters were happy to make that kinda deal. There was no money changin' hands plus it gave the dudes in the east a chance to see what we far out cats in the west were up to. Most booths go for between 4 and 5 hundred bucks, even in them days. That's when gas was 28 cents a gallon so gettin' to a show was no big thing.

Above: This sketch was probably penned after the Road Agent was some way along but the antennas never made it to the finished car. Left: Ed poses for the camera with his innovative upside down Corvair. ***Photos courtesy of Dirty Doug (above) and Petersen.***

Above: The antennas made it thus far but what you can't see is the shirt drying rack covered in dust. Right: Dick Cook was responsible for the Agent's tube frame. Note the side mounted gas tank and battery. *Photos courtesy of Doug Pizac and the Bob Hegge Collection (above) and Petersen.*

Left: Another bubble top, this one pink and petal shaped, from Acry Plastics. Below: Ed tries it on for size. Note the final shape of the side pods, one of which is missing. Front axle was '37 Ford suspended by a single VW torsion bar while the steering gear was from a '58 Austin. *Photos courtesy of Petersen.*

Left: According to Ed, Doug, who usually painted Ed's cars, suggested they try the new candy apple paint and Larry Watson to apply it. A young Larry is seen here laying it on. Below: When it was finished, Larry had this lovely lady pose in one of his plastic metalflake skirts. *Photos courtesy of Larry Watson.*

ROAD AGENT
Built by
ED "BIG DADDY" ROTH
Corvair Powered - Fiberglas Bodied
ROAD AGENT

Above: Finished early in 1964 and shot for the cover of the April issue of *Rod & Custom.* Above right: Show promoters like Fred Caldwell were happy to use Ed's cars to attract the crowds. *Photos courtesy of Ed (above), Fred Caldwell (poster), and Bob Lewinski who took this shot at a show in 1965.*

ROTH

Roth's daily driver was this tangerine orange '55 Chevy fitted with a fiberglass front end disguising a 390ci Ford and C-4 auto—Ed never liked the GM box. An Edelbrock hood scoop, radiused rear wheel wells and cheater slicks mounted on Astro wheels completed the competition look. ***Photos courtesy of Darrell Zipp (top left), George Schreiber (left), Ed Newton (below), and Ed (bottom left).***

Orbitron

At about the same time, 1964 or so, I got this brilliant idea ta make this car with three big headlights so that when the three primary-colored lights hit on the road they would be a white beam. "Orbitron" was a failure at the shows. I believe it was because we covered that shiny chrome Chevy engine up with a hood. It was a great lesson in design for me. Never cover up the engine unless ya can serve a worthy purpose.

I shoulda named it the Titanic because it was like tryin' to hold onto this giant sinkin' ship. All the other cars I'd built went over big at the shows. I had a formula but I didn't have the brains to figure it out then. I do now but I ain't gonna tell you.

Seems like I used the engine from my '55 Chevy—it was layin' around so I used it. I also used a straight front axle and a stock width axle for the rear. I can remember

Above: June 1964 saw work start on the Orbitron which, during the process of construction, sadly lost something from Newt's original artwork (top right). ***Photos courtesy of Griff Borgeson (above), Ed Newton (below, page 61 top left and bottom)), and Dirty Doug (artwork).***

"Roth's
Orbitron"

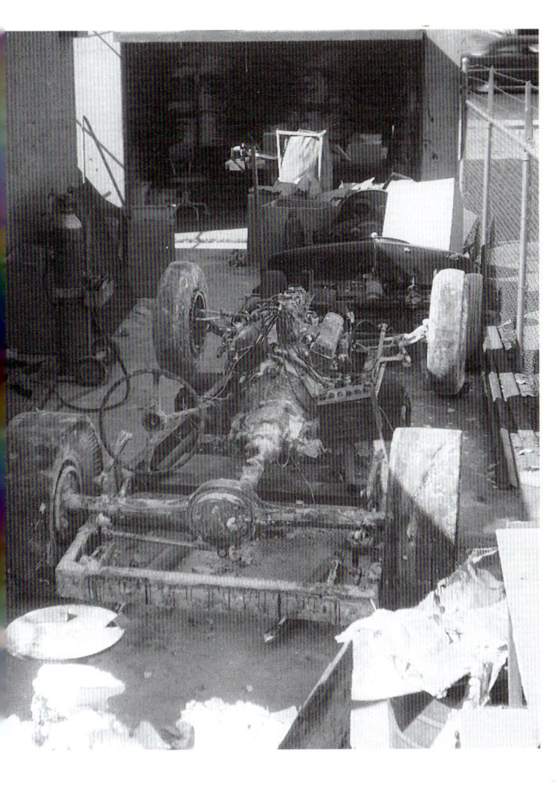

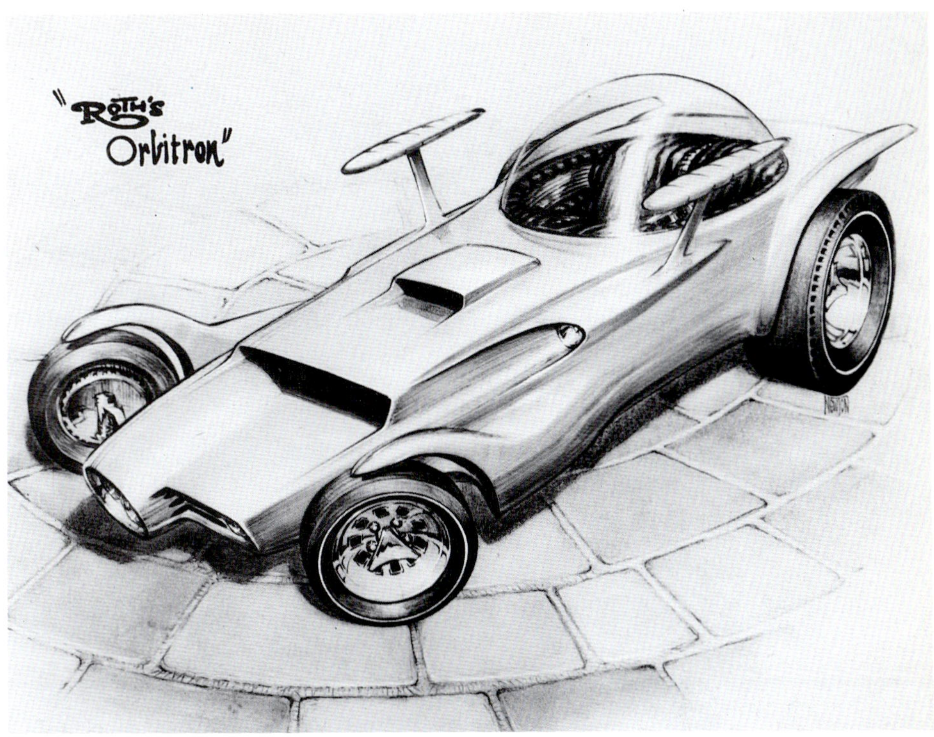

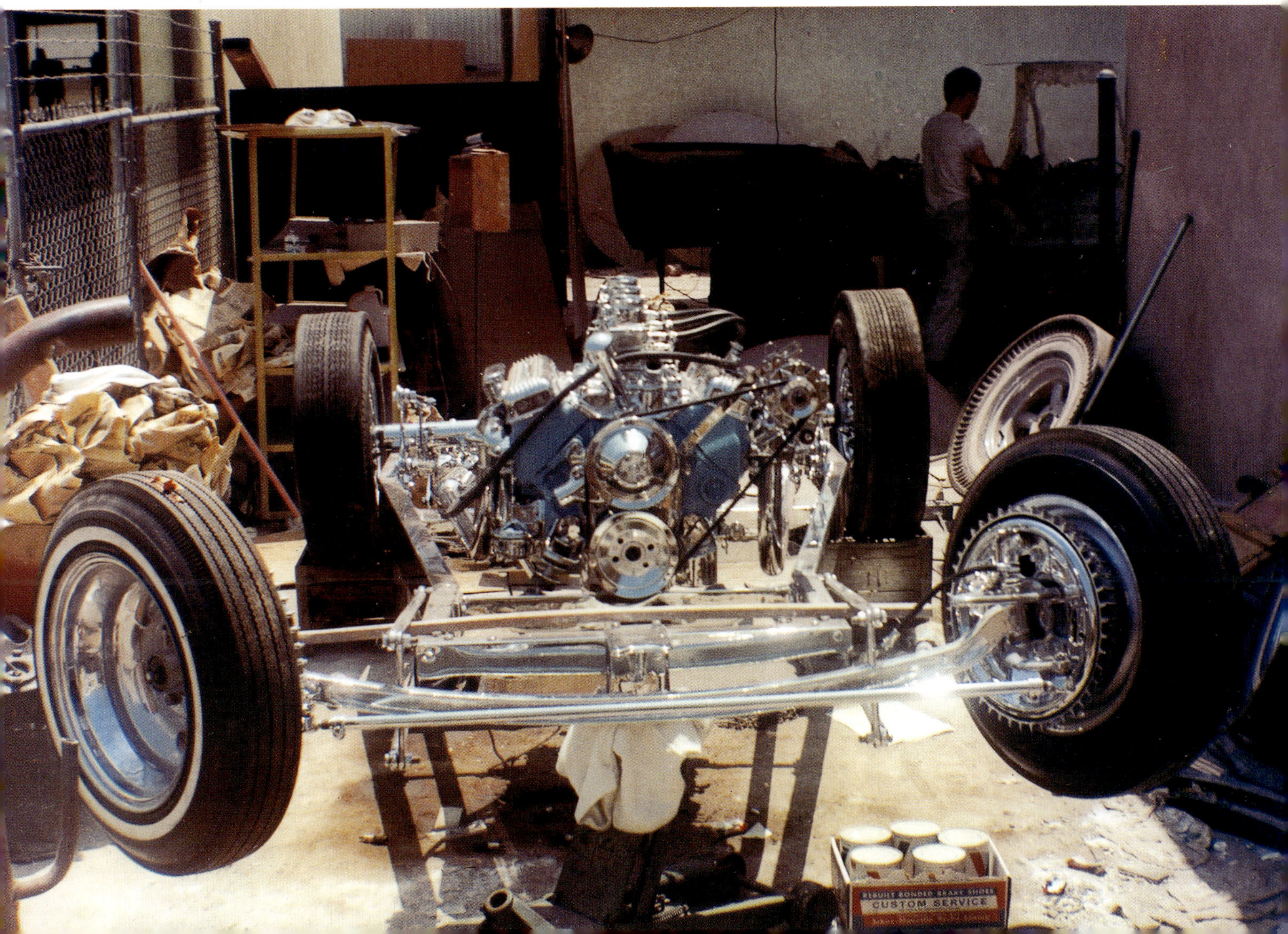
REBUILT BONDED BRAKE SHOES
CUSTOM SERVICE

Above: Resplendent in its Larry Watson-applied candy blue, Orbitron is seen here at Revell, presumably for measuring for a model. Right: Where it all went wrong, according to Ed, was in covering that shiny Chevy which came from his '55. *Photos courtesy of Darrell Zipp (above), and Ed (right).*

it was 1964 because Revell was houndin' me for more cars to make models of & I was humpin' my booty to get it over to Revell's HQ in Santa Monica, Calif. for them to measure when The Beatles appeared on the Ed Sullivan Show & all model sales stopped. Guys got guitars instead o' cars. Revell's sales fell off so bad that they almost went bust. The monster kits stopped too! It was the end of the model scene & the end of those big royalty checks.

I sold the Orbitron to some dude in Texas.

Top left: Seldom seen interior shot shows TV, Cragar wheel, Moon gas pedal, Hurst stick and Dixco tach. Bottom left: Not happy with the car, Ed had Larry Watson repaint it in pearl blue. Above: Newt's little joke turned out prophetic as the Orbitron disappeared in the mid sixties, however, just as we were putting this book to bed we got a call from Mike Lowe of El Paso, Texas, who bought it in 1973 and drove it to school before selling it. At the time of writing, knowing where it was, Mike was trying to buy it back. ***Photos courtesy of Darrell Zipp (top left), Petersen (bottom left and above left) and Larry Watson (left).***

Dirty Doug

Soon after Roth debuted the Outlaw at a show in a Huntington Park parking lot, I stopped in his teeny shop in Maywood and began to hang out on a regular basis. I'd do small errands for BDR and he'd let me mess around with the plaster to see how dirty I could get. I used to try and be a lady's man when I had my custom cars and I had all these suits that I'd wear to impress the girls. I'm not a wasteful person so I'd wear these suits to work in and it cracked Roth up so much that he started calling me "Dirty" Doug.

In the summer of 1960, BDR started traveling to car shows back east with the Outlaw but had started work on the Beatnik Bandit. The frame was chopped and some of the plaster was on it and while he was gone I worked on it. Personally, I think that he chickened out and passed the buck to me.

When we worked together I usually mixed the plaster and he applied it to the car and gave it the rough shape with his teeny little plasterer's trowel. I mixed the plaster with a broken yard stick. In the summer it set up quite rapidly so by the time he applied one bucket of plaster I was mixing another. So off he goes on his trip and I'm there by myself with the challenge to make this rough-looking thing as smooth as a showroom car. To speed things up I start to mix the plaster with my hands. The bucket is elbow deep and it's abrasive so my skin starts to wear away. I change hands. It starts to build up on both arms because I didn't wash them after every application. Soon my arms look like I have plaster casts and my fingers don't move so good. It was getting dark and I figured I'd wait to clean up at my house. When I got home my mom thought I was in an accident. My evening was spent trying to remove the plaster which hardened and bonded to every hair on my arms. I ripped most of my hairs out of my arms and slept with a long sleeve shirt that night but it didn't help much.

Shoes? Forget it. I had a pair of army surplus boots I'd wear that had a crust of fiberglass so heavy I could hardly walk in them. My pants were crusty with resin, and BDR had to buy a tuxedo to slip over his Levi's at the shows but he still reeked of fiberglass.

I had to block up the shortened Oldsmobile frame because the weight of the plaster caused it to settle and then the fenders would crack as they were built up from the top of the tires and not the frame like the rest of the body.

I bought one and a half tons of plaster during the construction. It seemed like 20 percent went on the floor as a low wall all around the outside of the body.

When he got back we primered the plaster and sanded it to a smooth finish. The mold we made weighed almost a half ton. BDR wasn't into molds at the time, but the finished body came out lightweight.

Building bodies is basically very simple. After the plaster is in finished form, I'd cover the plaster with three or four layers of resin & cloth, turn over the body and knock out the plaster and then sand the outer surface to a smooth finish using primer and filler.

Above: Dapper Doug tries on the Road Agent for size before standing back and operating the crowd-pleasing remote control. Bottom right: With the Orbitron, he told one onlooker that the three lights shined on to a white background would produce a white beam. *Photos courtesy of Dirty Doug.*

I would usually work on the bodies on 50-gallon oil drums while the parts were off getting chrome plated. It took six months for the plating in those days. That gave me time to finish the paint.

Paint. It was the good old days, remember? If I made a mistake or a fly landed in the paint I'd just rub it out. Later on Roth got Watson to paint the cars because Larry had a paint booth and he was coming up with candy colors that couldn't be rubbed. Martinez did the upholstery and Model Plating the chrome.

I did everything that a good gopher does: Drove to car shows, sold shirts, swept the yard along with his kids and tried to keep the floor clean, which was impossible. The plaster built up real fast because the mistakes ended up on the floor with the drippings from the plaster bucket. After every car was done we'd clean the floor and find screwdrivers and other missing tools.

BDR always had a lot of tools but his favorite was the torch. He used it as a saw, hammer, wrench and stove. When he'd cut a piece of metal for a bracket or start chopping a frame, the slag buildup would make the cut real stubborn so he'd whack the piece until it fell apart. It was easier than getting a hammer. Naturally it would bend the torch to match a pretzel. Luckily Harris Torch Repair was next door and the damage was easily repaired.

BDR loved to test people. In these later years I realized that he just likes new ideas and isn't afraid to try something new. I can remember when the Outlaw came back from the show circuit and it was wasted. It had nicks and scratches all over it. I guess BDR got a cash settlement from the promoter to fix the damage and I was elected to paint it. BDR just shrugged his shoulders when I asked him what color.

There was this new stuff called metalflake that Junior was messing with so I got a couple of pounds of green flake and started experimenting. Instead of using clear lacquer to build up the clear coat I used clear fiberglass and it never shrank.

BDR had five sons that he made hang out at the shop to keep them off the streets and out of trouble after school. They'd be all over the shop after they got through sweeping and sometimes I felt like the question-and-answer man. When they weren't getting into mischief they'd come over to my space and start bombarding me with questions.

"What are you doing Dirty Doug?" was the most common. I came up with a lot of different answers that left them giggling. That solved my relationship with them but didn't cure their curiosity. When BDR brought home the casket for the Druid Princess, I recall they had a contest to determine who fit it the best. There was no bounds to what they would venture into while BDR was off to a show. All of BDR's stuff started in gear and this lead to many unexplained broken fences and tire tracks going straight up the walls.

I remember starting the Mysterion up at the Sports Arena. It was the first time I'd ever demonstrated one of the cars, (BDR was off eating a hot dog), and I had one of the trannies in drive and it started to take off on me. I switched it off quickly. Frightening!

He would make the car start and raise the top or turn on the engine and lights by remote control. A lot of the cars had demonstration tapes and he would use country and western singers like Tex Ritter because he said that people understood them better. It would be nice to have some of those tapes today.

There were always a lot of people coming around the shop like bikers, hot rodders, show promoters, and designers like Joe Henning and Ed Newton, all making deals and putting in their own two cents worth. Yet it was up to BDR to make the final decision as to what the public wanted to see. Sometimes the changes that we made in the plaster were ridiculous. I'd have a plaster cast almost ready for the fiberglass cover and he'd come along and chop off a whole section that had taken me weeks to perfect.

After he'd chopped off the section that offended his eye, with this old, beat-up keyhole saw his dad gave him, he'd stick in several metal coat hangers and start rebuilding to his new vision. Then he'd call in Newt' or one of the other picture drawers and have them make a nice drawing of what he had so that he could get it to the car show promoters for some advance publicity.

The trash cans were a feast for me. All the sketches that Newt' and the other guys made hit the waste basket and I had a nightly trash can search and dug up a lot of the real neat "never before seen" sketches. I even dug one of the cars out of the trash: The Wishbone. BDR threw it away because he didn't think it was cool. I liked it for a lot of reasons. The narrowed rear end and double bubble were "Stoked". So he sold it to me for $75.

Big Daddy can't draw cars but he taught me a great lesson: You must build the car first and then have the artist come in to draw it. When I did the Orbitron, it looked pretty gross in person but when it was drawn on paper it looked pretty good.

The Orbitron had red, blue and green lights. At the Sports Arena a guy came up and asked me why the three colors and I says, "You focus red, blue, and green on a white background and you get a white beam." I said this and his jaw dropped to his chest in amazement. ❐

Surfite

I love junkyards! I mean I love to browse around in the wrecked dinosaurs & figure out why we Americans buy those big tuna boats. Also love to figure out where the weak points are & why designers do some o' those stupid things like build the center of gravity so high up to make the cars top heavy (all except the Fiero of course). I've loved junk yards from the first time that I ever went to one in the late forties. In those days ya hadda wait by the counter 'til the mechanic went back & unscrewed your part. In the '60s those "Pick-A-Part" yards let ya in & we could unscrew our own parts. I go in just for the kicks of it.

Left: Newt's original 1961 Art Center sketch for what became the Surfite. Above: Another big mess of plaster grows atop an Austin Mini Cooper chassis. Below: Doug's probably there, you just can't see him for dust. *Photos courtesy of Ed Newton.*

It was on one of those little escapades (about '62) that I spotted this little car that looked like the engine was in the trunk. I lifted the trunk lid & surprise, no engine. "Could the engine be under that snub nose?" I asked myself. There it was fer sure! It was mounted sideways. I looked at the brand name of the car & it said, "Mini Cooper". It was "Made in England" by Austin.

The junkie loaded it onto my trailer & Dick Cook took the body off & drove it around. It was definitely cute. While Ed Newton was workin' at the shop he sketched this cut little futuristic bomb. With a few changes, surfboard, etc., we were able to make it a genuine surfer's car & I called it the "Surfite".

Above: Another Dick Cook chassis this time housing a sidewinder 1269cc Austin Mini Cooper engine chromed as usual by Model Plating and fitted with, in Newt's opinion, oh-too-skinny Astro wheels. Left: Salvaged from the Revell trash can was this shot of the newly finished Surfite at Revell for measuring. *Photos courtesy of Ed Newton (above), and Darrell Zipp (left).*

I wasn't gonna cut loose of this story but I was in a movie. Well, I wasn't but the Surfite was. They made all the beach movies at Sportsman's Cove in Malibu & I got wind of this new movie Annette Funicello was makin' & she's my all time heart throb so I drove down & parked the Surfite next to Buster Keaton's dressing room, pretendin' to be one of the cast. If your hip to movie sets there's always good food around. I had it made.

Buster Keaton was an old time comic. He & I hit it off pretty good so when the honchos came around I'd start rappin' with Keaton. The reason I wanted the Surfite in "Beach Blanket Bingo" was for the advertising value. If I could advertise it as being in the movie I'd be able to collect $1,500 per show rather than the normal $200. So the movie gets made & out on the road but I can't spot the Surfite. Oh Man! What to do? By coincidence there's an old "Finkster" out in Hollywood who also happens to be a Surfite groupie. He comes to the show & tells me how great the Surfite looks on the screen. Then he pulls out these photos where Annette is walking in front of the Surfite for a split second. I'm safe. It was in the movie for a split second but it was definitely there.

Below: Newt's 1964 sketch for the Surfite clearly shows the wide, floatation-type tires he wanted. Right: Albeit for a split second only, the Surfite nevertheless made an appearance in Beach Blanket Bingo. *Artwork courtesy of Ed Newton, Photo (right) taken at Ray Farhner's Denver show about 1966 courtesy of Roger Kilborn.*

Ed Newton

Harry Costa's San Mateo car show was coming up and I tried to book it. Well, all of a sudden I find out Roth has the exclusive to debut the Road Agent. So I waited until the first day of the show and went down there. I had some photographs of some of the work that I had done. I caught him at a slow time and said maybe we could paint together. He told me to put the photos away, he didn't want other people seeing those things, saying, "You gonna wreck my business."

I had photos of an 1/8-scale clay model of something I had done at Art Center and he was more impressed with that than the shirts I had done. He kept at me until he finally made me an offer I couldn't refuse and so I moved down to Maywood.

He had purchased the rights to do the Surfite but one of the first things he wanted me to do was finish the Orbitron because he had it in such a state that he was unhappy with it. What I did first was make an idealistic rendering and it could have been a beautifully proportioned vehicle if we could have started from scratch. As it was, the Orbitron came out rather strange because it was conceived at a time when Ed was getting tired of having to do eveything himself, and I'm not taking anything away from Doug, Doug did an awful lot of the work, but as far as putting it together and making the decision of what went where, Ed did that.

From what he told me, he claimed that all he did with the Orbitron was take a couple of cardboard boxes and stick 'em out there and put four wheels around 'em and hold the steering wheel in his hand and he envisioned the car while he was sitting in the backyard where it grew as a plaster mass.

The Surfite was the first car on which I got out there and squooshed the plaster around and that's why it meant so much to me when it was put into model form by Revell. The only part that disappointed me was that I recommended go-kart style wide rims with flotation ATV-type tires but Ed had a deal with Kelly Tire and Cragar so we ended up with those skinny rims.

There were some little disappointments that happened after that. I had a tremendous concept for the Wishbone but when Ed took the drawings down to Royal Glazier and her husband at Revell to get approval, Royal looked at it and said, "That looks too much like the Road Agent." And Ed said, "But it doesn't even have a bubble top and the front end sticks way out." "Yeah, but it's still a wedge shape." So it was literally back to the drawing board.

Now that the idea of the wishbone was destroyed, which was all in the suspension, Ed didn't want to change the name, therefore I put this wishbone fork out front as a styling element which was a compromise we both disliked. Ed ended up hating the vehicle.

Revell never did make a model of the Wishbone. They had started on it, and it was in the mold-making stages when they pulled the plug. The plug was actually pulled on the Surfite so there were very few Surfite kits.

With the Bike Truck, Ed told me that he wanted to do a vehicle that would haul a motorcycle around. He had this little Honda that he had for a while, and I had a Triumph Bonneville show bike, and Fuller had a knucklehead Harley chopper, and we would both kinda laugh at him. So one day he goes out and buys a brand new XLCH and had me design this truck around the bike. There was a lot of consideration of the angles of the exhaust and various elements working together, and in the preliminary drawing I made sure that I did take the advice of Roth: "Don't do designs like Art Center. Show some of the mechanicals." Therefore I think the bike truck, as an exercise in exposing some of these mechanics, was fairly true to his original thought. He ended up putting more body work on it when he reworked the design. As he progressed into this particular vehicle he

---

According to Newt', this '31 Packard was destined for the Munsters TV show with a Rolls Royce grille and an Olds engine. However, the producers turned it down in favor of the Barris/Daniel effort. Also turned down was this OTT concept (right) for the Monkees. ***Artwork and photos courtesy of Ed Newton.***

would get different opinions so the original integrity was lost in the shuffle. However, he did end up with a very unique and dynamic vehicle and the only time that I really had to bury my head in my arms, was at the last minute he couldn't make the custom taillights that I had designed. He just went down and got some Plymouth Barracuda taillights and stuck them on the back.

But I was always trying to put these stylistic elements into the cars, like the Surfite we designed a wraparound red plexi taillight that worked stylistically, but because of time constraints, he never seemed to follow through. And I can understand because his obligations, and he tried to save a buck here or there, so it was a natural thing where that would happen.

Roth did an awful lot of his own stuff. He was usually the one who was responsible for going out there and beating it to death and squishing and slapping that plaster. I remember I made up a rear end for the Surfite and I thought it was beautiful and he came out and said, "Naw, that's not it yet. I don't like that." So I had to take a sledge hammer and knock the whole back end off the car.

What I did with the Druid Princess, I designed the large canister headlights and the radiator. Pretty much the car was envisioned in a fairly simple manner. It was a fairly conventional frame layout but instead of the body it was the coach and if I had been involved with designing the coach part of it, I would probably have stylized the coach. As it was, it looked like the old British Royal coaches even though it was fiberglass and plywood; it didn't have any real direction to it. You could take that thing off and turn it around 180 degrees and it would still look like it was going the right direction. But, yeah, I did the front end. I think Daniels was responsible for the fenders, those lazy oak fenders. I think Roth's idea certainly was the baby casket. And of course the concept of show cars grew into a more competitive, monetary, and egotistical battle from the various builders and in order to do this many of them would just create the illusion.

For instance the Road Agent was a fantastic concept of taking a Corvair engine and putting it in front of the axle except in order to make it work he had to turn the transmission upside down and the thing would leak out the top, which was the bottom. Still, it was the effort and he had the guts to do those things and that's what counted. Why I like the Road Agent so much is because it really was a technological breakthrough. He was always moving towards the innovative end of the spectrum. I'm not saying he broke any ground that hadn't been broken decades before, but in the culture we were in, the hot rod culture, everybody was following whoever was hot at the time. Roth was not following. He did things in the hot rod field that other hot rodders were afraid to do or didn't even consider, so that's where he shined. Most of his cars represented an attempt to break new ground, consequently, he would find out in the long haul that some of the grief that he went through earlier could be avoided by a couple of simple concepts. One concept was downsizing.

He was probably one of the first customizers or rod builders to really adopt using everything smaller. The very idea of going from the Outlaw, Beatnik Bandit and Mysterion— here's a huge car with twin engines and all that—down eventually to motorcycles, three-wheelers, and then these little itty bitty things like Rubber Ducky and the whatever. Now he's so proud of his little wagons. You can't get much smaller than that.

Roth and I were bidding on the various TV shows and the cars that were being made at the time. I still have some artwork that I did for the Munsters Coach which we lost because he made me design a car that was too conservative. He had an old Packard in the back yard and he wanted to put a Cadillac engine in it or some such thing and put a Rolls-Royce grille on it. So I drew one up and I submitted it and we got bumped by Tom Daniels' Barris Munster Coach. So I told Roth that the next time we had an opportunity, let me do something really wild, So the Monkees came up and Barris gave his version and [Dean] Jeffries gave his version and Roth let me do this four-wheeled, four-passenger vehicle that was like an electric guitar. It was the shape of a guitar but it looked like a hot rod and this was quite an extensive proposal. Well, they ended up doing this modified GTO that Jeffries had proposed and they were going for budget and we lost out again because we had over-engineered this thing. Unfortunately, we were never able to come to a happy medium and, of course, since I never did the actual estimate, that was left to Roth. He was the one who put a dollar value on it, and a lot of this might have had to do with the pricing on it. But we did have our opportunities to try to get these TV contracts. ❐

Wishbone

I guess the statement I wanted to make with the "Wishbone" was that rear-engined small cars were "where it's at." I picked a VW engine 'cause in the late-sixties they were comin' up with some tough engines. I narrowed the rear axles myself & used a newfangled front end that was half Ford and half VW. The front wheels were spokes 'cause they were real popular at the drag strip & easy to get.

The Wishbone was one car that I built that was such a sad piece when it was done that I decided to throw it away. It was so nerdy! Absolutely one big mistake! I got the saw & cut it in little pieces & took the frame, cut it in

Above: Newt's original concept sketch for the Wishbone was rejected by Royal Glazier of Revell saying it looked too much like the Road Agent. Below: Ed gets to grips with the Wishbone while the Surfite progresses in the background. Right: Those are some serious chassis rails waiting to be cut to length. ***Artwork courtesy of Ed newton. Photos courtesy of Tony Thacker (below), Ed Newton (right), and Dirty Doug (inset).***

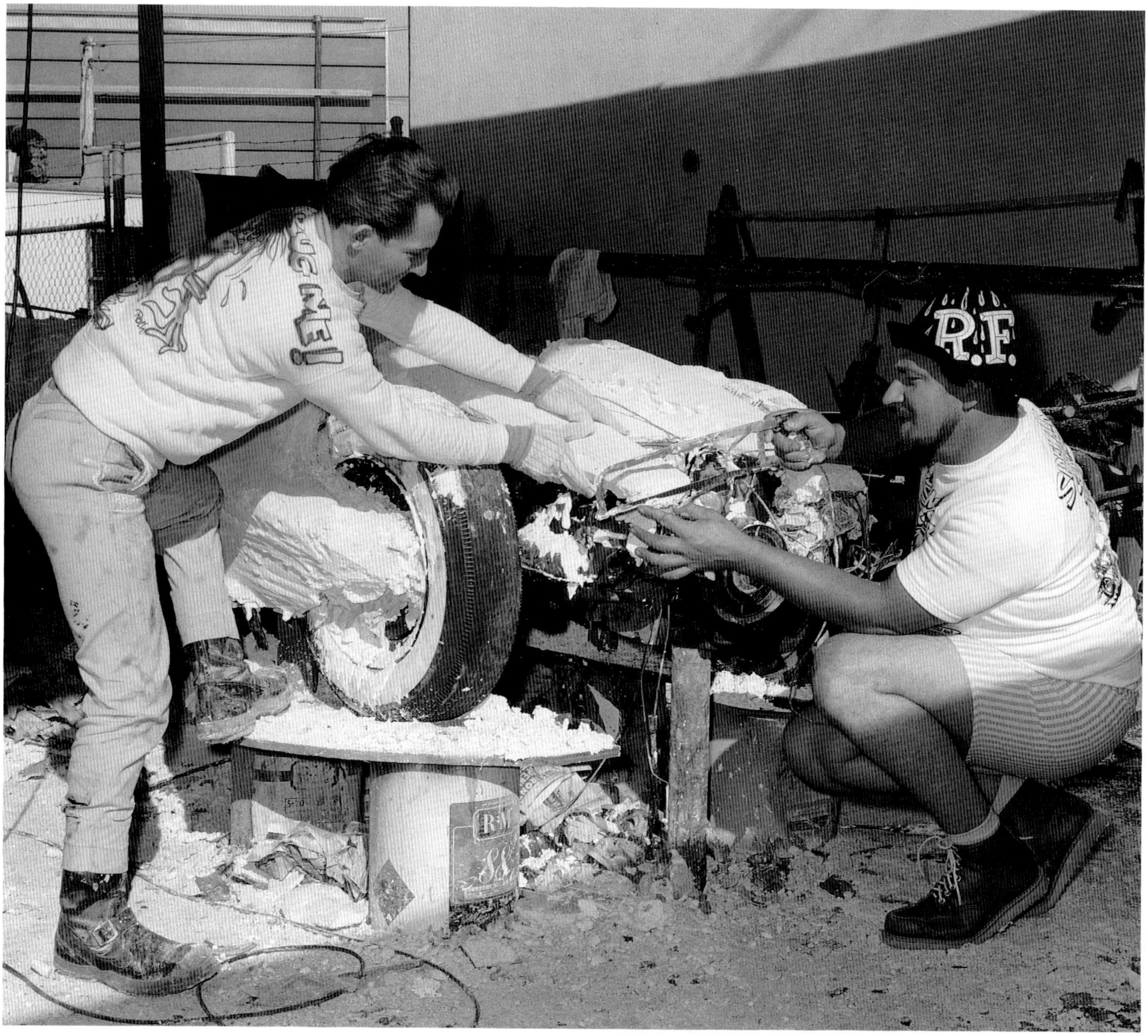

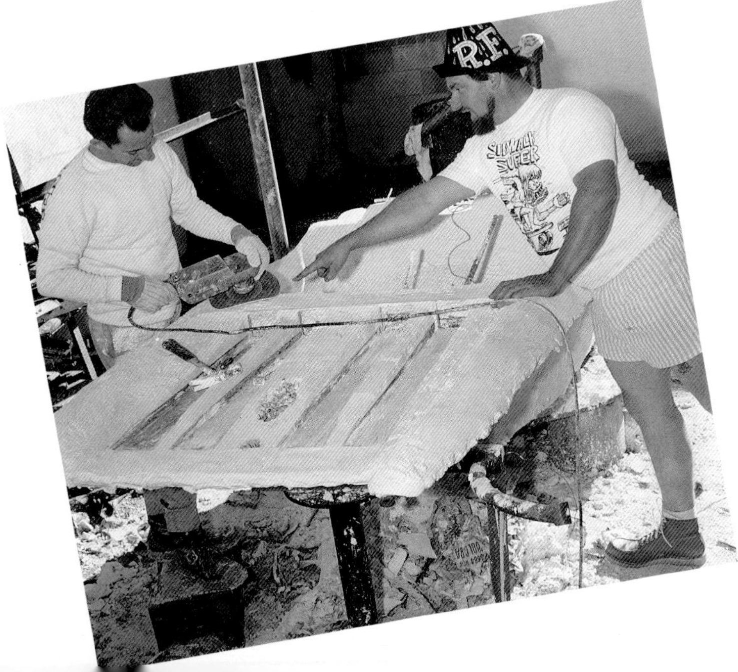

two or three pieces & put it in the trash cans. Dirty Doug came to me & was concerned about the futility of that act. He wanted it! "Okay. One stipulation," I told him. "You can have it if ya promise not to ever put it together again." I never wanted to have anyone see how klutzy I could design stuff. So I thought it was gone into DD's junk pile. Lost forever. Wrong.

Through a series of trades (DD's famous for trading!) this dude in Bell ends up with it. His motto musta been, "Ya can never get enough o' what ya don't need!" The next thing y'know the Wishbone was at the shows with my name on it! Barf.

Left: BDR and Doug ham it up for the cameras; however, Doug remembers that Ed would often hack off an offending section. With Revell already skeptical, compromising must have been hard to deal with for all concerned. *Photos courtesy of Petersen.*

Top: This shot puts things in a time frame showing both the metalflake Outlaw and the almost finished Rat Fink. Above: Doug shows off what's signposted one time only as the Electra at the Los Angeles Convention Center in 1970. Bottom: The real reason for this shot is that it shows Ed's Honda 50 in the background. *Photos courtesy of Doug.*

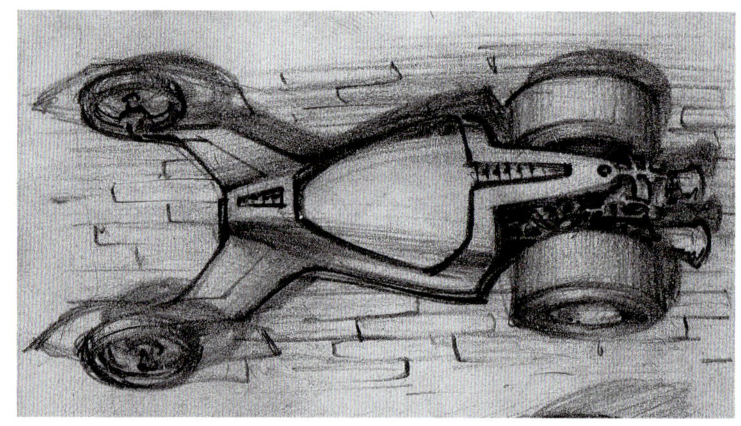

Right: According to Newt' this was one of three after-the-fact, what-if sketches exploring the Wishbone concept. Below: Darryl Roth proves it runs as he blats the rebuilt Wishbone down the street for Geoff Carter of *Street Rodder. Artwork courtesy of Griff Borgeson. Photo courtesy of Street Rodder.*

Left: Dirty Doug pieces together the split-Wishbone in Darryl's garage. If this were color you could still see the original paint. Below: Whichever way you look at it, the Wishbone was radical for its time. *Photos courtesy of Street Rodder.*

Above: Ed gets the keys to a brand new '64 Plymouth Super Stock. Left: If Ed's antics to the right were anything to go by, it was no wonder the Plymouth was soon up for sale. Right: A page from Pete Millar's Dragtoons. ***Photos courtesy of Ed (above), George Schreiber (left), and Pete Millar (right).***

B.D.R. GOES DRAGGIN'

WE RECENTLY FOLLOWED BIG DADDY ROTH, THE MAGIC DRAG-ON, AS HE RAN HIS PLYMOUTH THROUGH THE LIGHTS AT LIONS DRAG STRIP AT LONG BEACH. HE MANAGED TO KNOCK OUT ONE LIGHT AND STUN THE OTHER. NO, REALLY HE DID PRETTY FAIR IN THE E.T. DEPT., BUT UNFORTUNATELY HE WAS GOING THE WRONG WAY (SEE PHOTO). HE WIPED OUT 4 DRAGSTERS, 2 B GAS COUPES, ONE DOG WITH FLEAS, AND A DESTROYER ESCORT BEACHED IN A NEAR-BY VACANT LOT. HO-HUM . . . MAYBE SOME OTHER DAY!

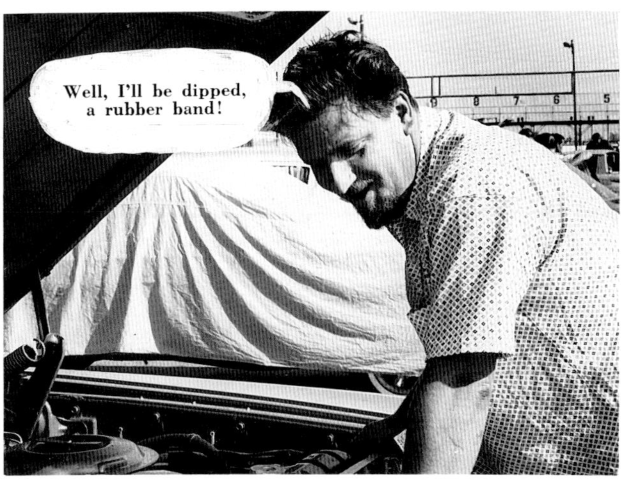

WE PLAN TO FOLLOW B.D.R. ON OTHER ADVENTURES, I.E. B.D.R. GOES SKY DIVING, PULLS A CAPER, RUNS NAKED, DANCES WITH A MAN-EATING PROBOSCUS MONKEY! . . . BE WITH US FOR MORE STOMACH TURNING, SICKENING, OFT' THRILLING ADVENTURES OF B.D.R.

Fang

'HALFBREED HANNA', THE RACECAR BUILDER'S RACECAR BUILDER, DOES THE DEED FOR BIG DADDY ROTH, WITH THE 'BUSHMASTER' WHEELIN' IT!!

In '64 Revell was gettin' ready to make a series of new models & part of the contract was to give me two new cars. Plymouth had just come out with the new street rod cars. These were real go-gettin' 11-second machines. Hardly any were available 'cause guys back east wuz scoopin' 'em up as fast as they came off the assembly line so I told Revell that I wanted two of them instead of royalties. A stocker for my wife to drive & a "Hottie" for me to drive. And Ya, I hadda get into racin'. It wasn't good enough that I was buildin' a show car a year, I hadda try & race the guys at the track too.

There was a lotta royalty checks comin' in & even in them days "Uncle Sammy" was takin' his fair share. I don't remember exactly where George "The Bushmaster" Schreiber came from. Actually a guy like Bush doesn't come from anywhere he's omnipresent. If your in the pits ya can hear him raving in the spectator area & vice a versa. He's the only guy I know who can out talk & outrace anyone else. The difference between Bush & any of the other mouthy bench racers is that he actually does what he says.

Bush is always clean & with all o' his wizecrackin' motormouthin' always polite & knowledgeable. So one day at the dragstrip he's carryin' on in the pit area about how this streamlining really does the trick. He wuz rantin' about Jocko's streamliner (still one of the best) cuttin' off a whole second of elapsed time with his trick sleek body. I was impressed & had the same theory & even though I loved the drag races I figured I was too stupid to build & drive a dragster. My racing was done from stop sign to stop sign in Bell & Maywood. Y'know, nuthin' big time like a rail job. So I walk up to George waitin' for him to breath so's I can get a word in edgewise. Y'see George has this habit o' not breathin' while he's talkin'. He keeps spewing out stuff till he turns red & somehow he inhales thru his ears 'cause the mouth is non stop. The crowd he's got around him consist of hot rodding's biggies: Mickey Thompson, Ed Iskenderian, etc. He's not just rappin' a line but he's wrenchin' on the motor of his Deuce roadster at the same time. Bush is one of two people I know that can carry on a highly complex conversation & do high tech wrenchin' at the same time—Fritz Voigt's the other. I scratch this note on the back o' one o' my business cards, "George if ya wanna do a streamliner meet me at the shop Monday morning," & I squeeze my arms thru the throng around him & put it in the carb in front o' him. Without pausin' his line o' rap he takes it & stuffs it in his wallet.

By the time he arrived at the shop I had a clay model ready. He was impressed. George gets impressed whenever his bald head turns red & his eyeballs get tears in 'em & he starts droolin'. I knew I had him. George had the guts & know how & I had a fat wallet. Partners. We decided to call it "Yellow Fang" cause Phyllis Diller was calling her husband "White Fang" in all her comedy gigs but the more I thought about it the more I realized that glas was not the answer because of the tremendous vibration & twisting a dragster goes thru when it leaves the line so we decided to have Tom Hanna in Chula Vista make the body outta aluminum. In them daze ('66) there was only a coupla guys that could do that type o' work. So we took the frame, which I think Jim Davis did, down to Tom & began those long trips down on weekends to

Far left: The aluminum work for the Fang was done at Tom Hanna's shop. That's Tom in the hat and Roth for the one and only time in the driver's seat. Above: Roth apparently couldn't stand to run a unlogoed car so he applied the logo with a magic marker. *Photos courtesy of George Schreiber (left) and Ed Newton (above).*

guide Tom's expert hand into making the clay model come to life. It took six months but it happened.

We went to a lot o' drag races & got a lotta prize money: Fontana, San Fernando, & Lions Drag Strip in Long Beach got to be our weekly hangouts. We rebuilt the blown Chrysler hemi engine weekly. Three runs wuz about all we could get from a set of main bearing inserts. The financial drain was killin' me. Then it happened at Carlsbad. George had suited up & I push started him to get that big fuel burnin' sucker started. After the engine lit up & the ground was shakin' like earthquake city I knew this could be a record breakin' run. We had been makin' earth shatterin' runs all afternoon & settin' the pill (fuel restricter for the injectors) at the maximum leanness for minimum elapsed time. We'd been doin' 7.3 second quarters & this time the engine was soundin' boss. When he did his burnout I pushed Fang back to the starting line. The Christmas tree started the countdown. Bush got a hole shot on his competitor. All the way down the asphalt strip that engine was cookin'. We had the right combo fer sure.

Then as Bush passed the timing light I sees this big bright red flash which I originally mistook for our yellow

parachute but it was red. It was dusk so I figure the sun's playin' tricks with my eyeballs. But No! The announcer says Fire! Fire! I jumped in the push truck & when I got to the end of the strip I was greeted with Bush fightin' the fire with an extinguisher & as more people came more extinguishers were emptied onto the raging inferno. Methanol is furious & the extinguishers were almost powerless. The fire was started from a hot engine that we ran too close to the edge. It just exploded! That was the end of my drag racing days. I also sold the Plymouth 'cause it was too fast & not dependable & used a ton o' gas to get to the grocery store.

At about this same time I'd been working on the California Cruiser trike and I continued on that project while George almost completely rebuilt Fang. I told George to take the racer & tour it wherever he wanted & if he ever sold it to split the money with me. He took it all over & built this special trailer for it with a big glass window. I know he went to Australia & Canada with it. We later sold it to Harrah's & he sold it to Don Garlit's drag race museum in Ocala, Florida, where it's on permanent display. It was costin' 250 bucks a week and a lotta sweat to get Fang to the strip and it was a relief for me to step outta that situation. The lessons were invaluable for my future projects. Nowadays "Bush" is racing a jet car. I like his truck & trailer better'n I like the car.

Left: The Groundshakers, literally, were Roth and the Bushmaster's first pit crew at Fang's Lions Drag Strip debut. Below: Resplendent in yellow the Steve Swaja-designed, Jim Davis-chassied Fang won many Top Eliminators and ran a best ever of 7.86 at 204 mph. ***Photos courtesy of Ed Newton (left) and Dave Peters.***

Right: After a successful sojourn in Australia, George, who now owned the Fang, had it repainted at the late great George Cerney's in Long Beach. The real reason for showing this shot taken at Half Moon Bay in 1969 was to show Roth's Chapel of Memories trailer which saved him from so much aggravation with the police while on the road. From left to right it's Frank Zinn, The Bushmaster, Jim Nicoll and Jesse Perkins. *Photos courtesy of George Schreiber.*

Above: 1967 and Roth's gang: from left to right, Dick Cook, Ed Fuller, Roteye (as George called Roth), and George Schreiber attempt the Mint 400 in this Dick Cook-built desert racer. According to Ed, they laughed at the competition, mostly V-Dubs, but failed to finish when the engine gave out. Right: Ed picked up this teeny weeny bike from Baggy Bagdaserian. It was too big for Ed but perfect for Dick Cook who could master anything mechanical. ***Photo courtesy of George Schreiber (above) and Mark Nelson.***

Bikes n' Trikes

It was in 1966 that I got hooked up with motorcycles. I had absolutely no desire to ride a bike. They were like horses. I was scared of 'em. But I figured it out like this. If guys were choppin' hogs there was somethin' I was missing. So I tools down to the Sheriff's auction in L.A. & put my bid in on 15 of the old scooters that were stripped down & used in the L.A. river bed to train new cops how to ride around pylons & how to lay the bike down at high speeds without hurtin' themselves. These bikes were so beat up that even the guys from the Tijuana police department didn't want 'em. All dented up & soundin' like Mack trucks. Most of 'em never even ran.

Surprise! Surprise! I end up with five of the buggers for about $250 a piece. I disassembles one & cleans it all up, put a sissy bar on it. That was 1966. I can remember 'cause I went to the "Hog" shop & bought me a brand new '66 "Shovelhead" engine for this wasted '65 Panhead and topped it off with a new 2-inch SU carburetor—I was king of the hill.

Even in them days, Harleys had this mystery about 'em. Nobody could figure out what made guys climb onto these ugly things & cruise around on 'em. I couldn't handle the fact that guys wuz buildin' these "Choppers" on their coffee tables in their front rooms. One of the biggest coffee table chopper builders was this chrome guy I mentioned before "Mando." His whole life was wrapped up in comin' home from the chrome shop & tinkerin' with his bike in the front room. His "old lady" was happy 'cause he was home with her & the kids. I think Mando started the chrome shop so's he could build a totally chrome bike for himself.

That's what freaked me out. Why would any "normal" person do such a crazy stunt? What was there about Harleys that hypnotized people? Like the saying goes, "If I gotta explain, you wouldn't understand!". Y'see there is no explanation. It's a spiritual thing that ya can

Left: Ed and Oink at the YTC, a sorta juvenile jail, in Chino, California. Right: Newt's concept sketch for the California Cruiser which depicts the twin head-lights seen on the early mock up below. They were later dropped in favor of the single pop up. *Photos courtesy of Ed (left and right), and Dan Woods (below).*

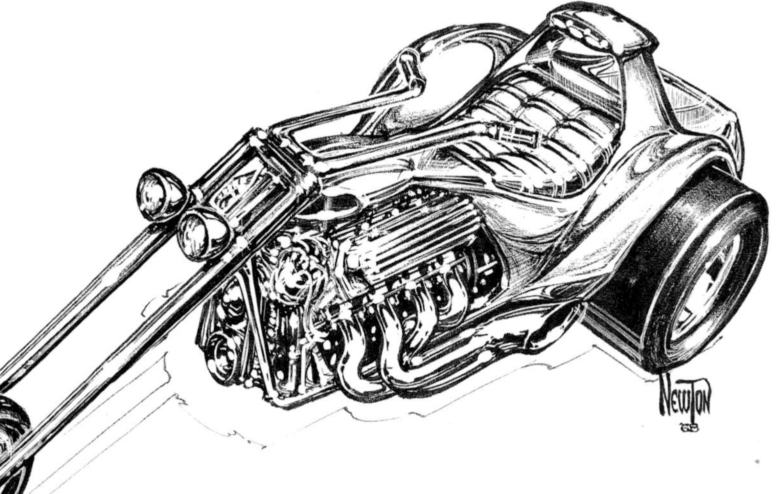

never learn unless ya take part! I decided to take part. I'll tell ya why. I wanted to find out why the public put down choppers & why such a totally American thing was so popular & why guys dedicated their lives to it. Ya see we can put down somethin' but if it gets to be popular, like hot rodding, there must be something totally "Free" about it. In other words the movement has'ta have class & it can only be done by minds born in a free society.

I started goin' on the AMA runs. The American Motorcycling Association had about 10 of these events for district 37 (So Cal) every summer. I tagged along. Those big "Hog" riders didn't really appreciate us chopper boys. They sorta figured choppin' a hog was sacrilegious. Ya we chopped 'em. We removed every unnecessary part of that bike that made it lighter & more of a screamer. On the other hand I'd ride with a bunch o' outlaw bikers (Y'know long hair, dope, choppers, etc.) just to get on the road. These guys were willin' to ride anytime, day or night. No organization, just ridin' & partyin'. There was doctors & lawyers in both groups although the wine & weed flowed more with the outlaws. They were called one percenters 'cause they was in the minority.

I took my camera on all these runs. I had pictures of a lotta choppers & full dressers so I called the owner of *Hot Rod Magazine* & sez, "Pete, your missin' out on a lotta good machines & I think ya oughta print some of these high tech choppers in your book." He got so irritated. "I'll never print that trash," was his final reply! Click!

The way I came home & printed my own *Choppers Magazine* for three years is history, but the next thing I know is Revell calls me in & politely informs me that due to an article in some big time magazine in which they quoted me as being the "supply sergeant for the Hell's Angels," they would have to cancel my contract if I didn't straighten up. Wow! Can ya dig it? All those goodtime royalties down the tube. Gulp! My reply was,

Above: January 1970 and Ed visits his friend Dean Moon whose rickshaw that is in the background. The original Cruiser was Buick F-85 powered, however, production "Cobra" trikes were Mustang V8 powered such as the one left owned by Neal Redfearn. ***Photos courtesy of Dean Moon Jr. (above), and Terry Thompson (left).***

"How can anything so traditionally American be harmful?"

I decided to continue the bike scene based on the fact that those early choppers were built with the same, if not more, tender loving care that the hot rods were built with. In retrospect, I made the correct decision. If I'd have sold out to the "money thing" I'd be a rich but sad dude.This way, I understand the Harley mystique which is more than bucks could ever buy for me!

So there I was, nose to the wind, headin' towards Bakersfield up the Ridge Route out of Los Angeles. It's mountain after mountain. My bike was fast—at least it scared the snot right outa me at times—I was in tall cotton. It was dusk. My favorite game was to go the speed limit 'til some smart aleck would try to pass me then race him & pull away from 'em. It worked! It was a blast! Then it happened. The headlights appeared as before & just as he's about to pass I nailed the throttle with all it had. This turkey was stayin' up with me! Was my engine goin' sour? All kindsa stuff was rushin thru my brain. Actin' as nonchalant as possible I peeps to the left to see what is goin' on & this guy waves to me & floorboards it & leaves me like I was standin' still. It musta been close to 100 mph. I'll never forget that as long as I live. That new '66 Mustang did a number on me. I was so ticked off I got off at the next off ramp & headed home! I was sad for days.

But listen to this, as I was explaining the Ridge adventure to The Bushmaster he clammed up & I believe I saw a tear streamin' from his eyes (hard tellin' with his coke-bottle glasses). He says, "If we'da had Yellow Fang there we'da wiped him out." Wotta idea! A dragster for the streets. Only one problem. The cops! They'd bust me fer sure.

I got this idea. If I was to take one of the narrowed rear ends from the dragster & set it here & a motor & set it here (I put the motor on a milk crate) & put a Harley front end under the front I would have a legal dragster that couldn't be beat. Bush laughed so hard that he knocked the engine off the milk crate rollin' around the floor. That made me even more determined. Next day I was at the junkyard checkin' out Buick F-85 engines & trannys. Again, I dug the Chevy engines but the Buick automatic tranny was more trouble free & shifted into more positive gears—it weighed less too.

I came home & welded this nightmare together & I told Bush to take the dragster by himself 'cause I wanted to have some cheap thrills. Here was a machine that I could ride every day for 20 bucks a week & beat everything. The "California Cruiser" was born. Throttle response was phenomenal. It was fast with that 2.50:1 rear end & nothing could beat it, not even new Mustangs & I was a happy camper! Oink, I sold to my best friend Woody.

Looking back, I gotta tell ya the front forks of those V8 trikes gave us a fit. Y'can imagine settin' a small-block on a set o' forks designed for a big twin Harley. On the first attempt we used Glide forks but at a 40 degree rake they sorta bent & flexed instead of hydraulicin' so we'd put Sportster rear shock springs between the bottom leg & bottom triple tree to give the forks a boost. The whole picture was sick especially when ya think about one single spoke at a time supported the V8 engines. In 1985 Odis Smith of Bell Gardens came up with a home built sturdier fork that with the new Harley alloy rims really did the job of supportin' those big mills. I sighed a big relief.

I had good times on that bike. The car guys chased me away from the car digs & the bike guys refused to accept it as a bike. I went mostly on bike runs. I parked the bike in the shop & one morning I came in & I saw this tire track headin' right straight up the front wall in front of the bike. Someone had started it in gear & run it up the wall. 'Course none of my five sons copped to it. Musta been the neighbors!

Gotta admit I'm lucky to be alive! It was a dangerous machine. I made a slightly larger version for Ford Cobra engines. Several people died on 'em. They were top heavy & at speeds over 120 mph would almost rip the skin from your head. You couldn't close your mouth.

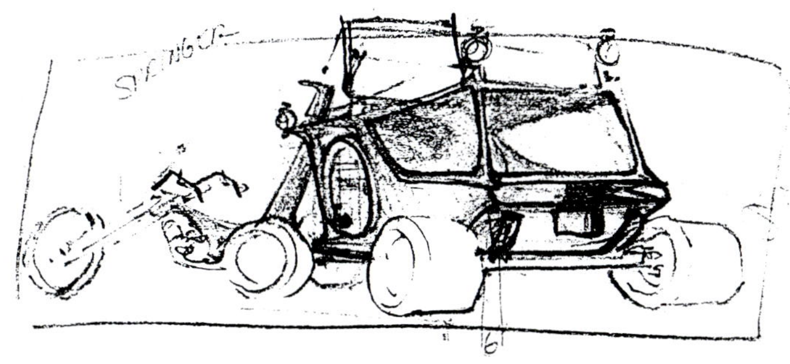

Above: Newt's original concept sketch for the Candy Wagon called for "two choppers pulling a wagon" complete with integral beer keg. Some sketches even depicted a machine gun mounted on the top of the wagon, presumably to stop kids from stealing candy. *Artwork courtesy of Ed Newton. Photo courtesy of Tony Thacker.*

Top: Roth and Tom McMullen conducted a trike test for McMullen's new bike magazine. Above: Two shots of the Candy Wagon, complete with candy, taken at Long Beach in 1968. Right: My favorite shot of Ed obviously enjoying himself on Slauson Avenue even though the wagon topped out at about 45 mph. ***Photos courtesy of Ed (top and right), and Jim Ballard (above).***

It was like ridin' in a space craft. It was time for something less drastic. Perhaps a Harley 45. It would surely work a lot better. Back to the Sheriff's auction.

Candy Wagon

I got the idea for the "Candy Wagon" at the Sheriff's bike auction in downtown L.A. These are the same underpowered trikes the cops use to give to the meter maids to slap the backs of the car tires (before parking meters) with a bag of white chalk. If the chalk was still

there when they came around again Bingo! Ticket! They had left-handed throttles so they could smack the chalk dust (on the end of a stick of course) on the tire. I bid 50 bucks for one & got it.

I built a trailer that I hauled the candy in for the kids at the parades. I had to stop 'cause the kids rushed in for the candy & spooked the horses The Candy Wagon was nuthin to write home about. Too slow. Back to the drawing board!

Everybody it seemed was building trikes, even Jim "Jake" Jacobs. The Mail Box had the entire running gear from a Crosley. Y'see the Crosley had no motor head. It was a mini Offenhauser & very trouble free & used in the 1/4 midgets in the CA area. Jake lost interest so I took the project over and finished it. The bike was neat but lemme tell ya it had no rear vision. It was scary. With that bike I learned a very important lesson & that is "Never cover your head with anything". Loss of sound & sight hinder motorcycle safety. That's why helmet laws do not appeal to me.

Druid Princess

I got a call from the motion picture studios to say that they were producing a new series called the "Addams Family." The producer wanted a car for this weird family to drive much like the one Barris had done for the "Munsters." In those years Barris was the one & only guy to build TV cars and whenever the studios'd call I'd always send 'em over to Barris because he had that stage presence to sell those 500 buck jobs for five grand. Anyway, I told 'em I'd take a crack at it & when I had it almost completely done I got a call from the studios that the series had been canceled. No more Hollywood for me, I was left holdin' the bag but finished it anyway & sent it out on tour to the eastern shows.

I remember tryin' to find a kids coffin for the trunk of the Princess (named by Robert Williams), so I went to this real exclusive, uptown coffin maker to buy one. The head honcho says "What'cha want this for?" I told him and he refused to sell me one 'cause he was afraid I was one o' them hippies tryin' to make a love wagon. So, as I'm walkin' outta the place this young kid (turns out to be the owner's kid) comes up to me & tells me to hand him 75 bucks. This kid is with the program! He knows who I am & that I'll do a bangup job on whatever I put down. He says, "Pull in the back alley & slow down." I

do as he says and as I slow down I feel this thud in back & there's this really farout kids coffin. That's where the gas tank & battery reside.

The fancy decorations were glued onto plywood & painted by Larry Watson. Those deco pieces come from a picture frame manufacturer in Hollywood. This dude named Skip Barret useta work for me & whenever I needed anything he'd know where to find it or he would make some phone calls & find it. I went over to the guy's shop & there was all these workers mixin' up this mixture of horse hoof glue & sawdust into this real doughboy type guck. It smelled pretty good but tasted terrible. Then they'd pour it into these floral-shaped molds. Some looked like little angels & others like these real swirly flowers. Y'know, just like ya see on the real expensive picture frames at downtown galleries. This guy hands me a catalog and I was flabbergasted at the great variety. As these little dough designs came outta the other end of the oven they were ready to apply on whatever. One look-see at the finished designs & I knew they were right for the Druid Princess decoration. I got me a coupla sacks full for 25 bucks & went home & started cuttin' & pastin".

Below: Though Tom Daniel also did some sketches of the Druid Princess, this one is the work of Newt'. Above: Jake, on the ground, and Woods in the chromed German helmet fool with the Princess while Ed, left, repairs the Road Agent after a tough time on the show circuit. ***Artwork courtesy of Ed Newton. Photos courtesy of Jake (color), and Ed (B/W).***

The finished product was a success. Later on the designs got kicked & scratched at the shows and it was badly in need of repair. My son Darryl bought it & I fiberglased over the whole car. It's on display in Europe now ('93) & goin' strong!

The engine & tranny for the Druid Princess came from the Dodge Corp. & of course, the blower has a carburetor in it so's it'd start easier at the shows & stuff. This car was almost totally built by myself, Jim Jacobs and Dirty Doug, with a strong assist from Dan Woods.

Left: Ed looks on as son Dennis waxes. Far left: After the Roth boys tried it on for size, the coffin housed the gas tank. Bottom left and right: Difficult to see but Watson's paint technique was called veiling as he used a special, Watson-designed three-headed gun. *Photos courtesy of Ed (top left), Pete Millar (top right), and Larry Watson (below).*

Capt. Pepi

About the same time I was really trippin' out on bikes. I went to Laidlaw's & got me a '66 Sportster & soon found out what bikes were all about. I realized that bikes'd soon play a major role in our country & decided to take a crack at predicting what El Caminos of the 21st century would look like. I wasn't too far wrong.

First I chain hoisted the Sportster to the ceiling & cut out cardboard templates for a plaster mold. I got me a junkyard even-fire Buick V6 engine & it fell into place with the odd looking body. Trouble was the stock Sportster on back. Thru a series of trades I got the Trumpet which sits on the back from a fellow show exhibitor Bob Aquistapace from Hollister, California. He traded me for the XL because he was tired of the trumpet he had built & shown for so many years.

The original name, "Captain Pepi's Motorcycle & Zeppelin Repair" came from Robert Williams' fruitful mind though it was later changed to "Mega Cycle" after a strong suggestion from car show promoters. After Mega Cycle came back from the eastern shows it was beat. Meanwhile, I had run outa room in my yard to store all these cars & my wife was already sore about the "no lawn" look in our backyard so I was lookin' for a safe place to store it after the show season was over. I was also, and this is a constant obsession, lookin' for a V-Dub engine for another car which turned out to be a trike. I hot footed it to the junk dealer & laid this deal out for him. I'd let him use the "Mega Cycle" in front of his junkyard for publicity for one year in trade for a V-Dub rear clip. "Done!" says the head honcho. At the end of the year I goes to pick up my beautiful Mega Cycle & this creep was lettin' his junkyard dogs use it for a doghouse. He never did pay me for the ripped seats. I sold it

Above: According to Newt', early 1966 concept sketches for the Megahauler, aka, Captain Pepi's Zeppelin & Motorcycle Repair, aka the Missing Link. Originally intended for hauling Roth's Harley XLCH until he purchased a show-winning Triumph. Left: Exposed mechanicals were appealing to Ed. *Artwork courtesy of Dirty Doug (above), and Ed Newton (left).*

Ed's compromise factor, or maybe the fact that none of his staff liked the bike truck, might have played a factor in the vehicle's diversion from Newt's original concept. Machine was Buick V6 powered with Corvair front end and rack and pinion steering. *Photos courtesy of Street Rodder.*

to Jim Brucker in about 1969 & he traded it to Harrah's.

When Bill Harrah died he had a collection of 800 or more cars. The head honchos decided to auction off most of the collection and it so happened that there were 10 or so cars of mine up there. So my son Darryl ended up with the Druid Princess & the Mega Cycle. He & I restored them with a giant assist from Pete Santini, who opened his shop to us & painted them when we massaged them with lotsa 320 sandpaper. Both were eventually sold to Jay Ohberg and are somewhere on the show circuit.

Above: According to Ed, the XLCH just didn't work and he didn't have time to build a bike so he traded Bob Acquistapace (right) for his show Triumph. *Photos courtesy of Roger Kilborn (above taken at Ray Farhner's show in Denver, Colorado), and Dan Woods (right).*

Dan Woods

I went to work for $3 or $2.75, something under five bucks an hour which was okay in 1966. When I got there he had this fifties Rolls Royce sedan and he had taken the motor out and was putting in an Oldsmobile engine. He'd had all the chrome done at Model Plating and he was going to make a hot rod out of it but he lost interest and traded it for a set of small-block Chevy valve covers. Back then there was at least a $1,000 worth of plating but that was typical of him, he didn't care.

If I remember right, when I got there, Ed already had the plywood body for the Druid Princess and there were four coil springs holding it up. There was some steel but there wasn't even frame rails. Newt's drawing was on the wall.

I was at Compton Junior College and had access to a machine shop so I made these nice little rings to hold the coil springs and brought 'em down and Ed was just dumbfounded that the things fit.

Dirty Doug was working on the body outside the shop whereas I was doing the chassis inside. Dirt did the whole body on that. It started out as a plywood box until they found a place that made all these things that go on picture frames to decorate it with.

There was a coffin on the back that he insisted be a real baby coffin. I had a sedan delivery that I was driving around and he asked me to go pick up this baby coffin. Now you can't get a coffin without a death certificate so we had go of all places to Watts and we had to do it at eleven or twelve o'clock at night and the Watts riots had just been a year and half before. So we go to this morturary and go in the back and meet this guy and we got cash money, like $200, and buy this little baby coffin. Jake and I put it in the back of my panel truck and we go to Harvey's, we head right to Harveys, we've got a coffin man, we're excited.

Ed wanted everything radiused and we'd file 'em and Roth was just dumbfounded. And when we took the first load of chrome to Model Plating they said, "Something's changed at Roth's, things don't look like they used to, they're detailed." It became a big deal around there. Ed had something in him that bugged him that it was too nice, he enjoyed it, he gave Jake and I the freedom to do this, but it wasn't Big Daddy Roth.

He wanted a blown motor and I was so proud of the linkage; it came out of the back of the blower and it looked like it was just a pivot point for a linkage to run the injector. I was just elated but Big Daddy comes out and scratches and says, "I want them to know there's a carburetor in there." And he took a drill and drilled four holes in the side of the blower so you could put a screwdriver in to adjust the carburetor.

Eventually we got the Princess running, but it had gotta go to Joe Perez' to get upholstered so Big Daddy says, "Well, let's just drive it over there." I says, "Great." So I jumped in this thing, there was no seat or nothing, and I got to Florence and I hit the light, it was one or two o'clock in the morning, and this lady pulls up next to me and she looks over and there's a baby coffin on this thing. We get to the next signal, and these drunks pull up and it was a scream.

Woods and Ed bleed the Road Agent's brakes during a post show circuit rebuild. *Photo courtesy of Dan Woods.*

Ed liked the Druid Princess because it was goofy and he loved that coffin on the back and the bizarre cherubs on the outside. We hauled it over to Watson and it got painted in a day and a half. I got Jim Babbs to make the radiator and Gilbert Metal Products made the lights but he made all that brass and stuff right there in the shop; outta tin cans if he'd had his way.

I think San Bernardino was the first show it was ever at and Jake and I unloaded it and Jake got in it and jumped on it and the motor quit and it never ran from that point on.

We were two-thirds into the Druid Princess when he got all excited about this bike carrying thing and we were into that when he got into the three wheelers. He was sure back then that in the next ten years every new car in America was going to have three wheels. It was just the way he thought.

With the bike rack we had to make one side of the rear end have a 14 inch half shaft while the other side was four and a half foot because everything was offset. I'd go order parts for this thing and people would go, "What?"

It needed taillights and one of the ugliest cars around at the time was the Valiant and Ed comes back with these Valiant taillights and Dirty Doug at that point just got totally frustrated, he couldn't handle it any more, and Jake was just totally amazed. I think that was when it stopped being fun. Besides, Ed was doing all this trike stuff, and I wasn't into it.

I think the way Ed saw it was that when he was building the Beatnik Bandit and the Road Agent it was fun and you could do whatever you wanted to do and nobody ever criticised the workmanship. Then, all of a sudden, when Jake and I got involved, things had to be nice, it took longer, his time was getting more involved and he just wanted to be involved with concepts and this trike happened in probably less than six weeks. They sculpted that body out of plaster using his bayonet and archaic equipment. One side was off from the other but it didn't matter, he was having fun. ❐

Darryl Roth

I found out that Brucker was selling all the cars to Harrah's and Big Daddy had the first right of refusal and I wanted to buy one of the cars and I told Big Daddy, "Hey, I've got the money, I want to buy one of the cars."

First he told me I couldn't afford it and then when I said, "Hey. I've got the money," he said, "You don't deserve it." That crushed me so I was bound and determined I was gonna get one of the cars.

I knew Dirty Doug had the Wishbone in his back yard and it was in pieces and I bought the pieces back. The front wheels were gone so I had Buchanan's make wheels for it. I think originally it had pneumatic bicycle wheels but I ended up putting a little heavier wheel on it and that was a real expensive part of the rebuild. And he'd cut the frame so we had to rechannel the frame.

Dirty Doug told me that Big Daddy hated the car so bad that he cut it in half and gave it to Dirty Doug for fifty dollars on the understanding that he would never put it back together again. The frame was chrome and it had held up pretty good and I just painted over the joint.

It had dual Ford carburetors on it and I remember it just used to foul up spark plugs but when *Street Rodder* came to photograph it and I said, "I'm gonna drive it," Geoff Carter said, "This thing runs?" I said, "Yeah. Let's go for a ride."

That feature just put me and Big Daddy on opposite ends of the world—big time. We had some other problems in between. When I finally sold that car that was the beginning of rectifying our problems. Actually I didn't sell it, I traded it to Harrah's for the Captain Pepi's.

I talked to Clyde Wade at Harrah's and he told me, "The car's pretty well trashed but we'll trade ya." I said, "Alright." That was the one I originally wanted to buy when I talked to Big Daddy and he said, "Well, you can't afford it. You don't deserve it." That pissed me off. I was going to get it, one way or another.

I remember cleaning it. It sat on the front row at Cars of the Stars and I remember cleaning it. And I don't know if Big Daddy knows this today. They had a light inside that car that shined up inside and I knocked the light down and burned the carpet and the upholstery and I remember that Big Daddy was flamin' and of course, I didn't do it. Nobody did it.

I really wanted that car because I liked sittin' in it. You had to be a little guy to get in it but you could sit in it an watch people and they couldn't see me.

When I restored the bike hauler, I think Big Daddy was upset that it looked better than it did the first time. He may have been upset that everybody was saying, "Wow, it looks so much better." When the guy from *Truckin' Magazine* came out to photograph it, Big Daddy walked up in his top hat and tails and handed the guy a tape and said, "All your answers are here, let's shoot the car and I'm leaving."

It was just like that and when the thing came out in the magazine, I got zero acknowledgement. It was at that point I figured out this is Big Daddy's car no matter what I do to it, it's always gonna be his car but in my own mind it looked better than what it had. It had matching

wheels, it had matching paint for the motorcycle and I thought it looked great. It was just an updated version.

When we did the Druid Princess, I knew how I wanted it painted, actually, my little girl always said that was her car and I asked her what color she wanted it painted and she said, "Purple." That's why I painted it purple—'cause my daughter wanted it painted purple.

So I told Santini, "Petey, I want to paint it purple." So I went by and looked at it and I thought it looked great. He was doing the fenders light and the body dark and he had this tint fade and I loved it.

So Big Daddy said he wanted to see it. So he went and looked at it and he said, "Hey, Pete, it's too dark. Let me do a little airbrushing on it and I'll make it look good." So Big Daddy went out there to do the airbrushing on it and he airbrushed stick men.

The next day, Santini calls me and says, "Hey, bro, your dad ruined the car." I said, "What are you talking about? No way."

I called Big Daddy and said, "What'd you do to the car?" And he said, "Well, I think it looks pretty good." And I said, "Well, I'll go out and look at it." And I went out to Santini's and it It really looked terrible and I was really upset. And I said to Pete, "What do you think about repainting it?" And he said, "You can't do that." And I said, "Hey, Pete, it's my car. I paid for it. Let's paint it." At the time, me and my dad were at one another's throats and I was just dumping money into this car and now we got a third rock head involved: Me, Santini and my dad.

I was so upset I said to Santini, "I'll come out tonight and we'll repaint it." So I went to Santini's shop after work and we completely repainted the car.

At five am my phone rings and it's Big Daddy and he says, "Hey I wanted to catch you before you went to work," he says, "What'd you think?" And I says, "About what?" He said, "'Bout the Druid Princess?" "Not much!" I replied. He says, "Well, what'd you do?" "Well, we changed it," and he said, "You repainted the whole car didn't you?" and I said, "Well, yeah" and he said, "Why'd you that?" "It was too dark and it needed lightening up." And he said, "I knew you were going to repaint it. Why'd you intentionally go there and change what I had done?"

"Because I didn't like it." That's one of the few times I've told my dad what was really on my mind. And there was a silence on the phone. I had never heard my dad have nothing to say. Then I realized what I'd said to my dad and I thought, wait a minute, I'm gonna have to figure out a way to clean this up. And I said, "Hey, those stick men looked terrible. When you said you were going out there to airbrush, I thought you were going to do a good job." Then I realized I was just digging a bigger hole. It looked terrible. "I thought you were going to do faces on the angles, I thought you were going to put strings on the harps. You went and did stick men."

After it was all done to my satisfaction and went in the show and Big Daddy was there and everybody was around and houndin' him for autographs, and he's doing the big smile and the tongue out and the tiltin' top hat and that was like, "Hey, you're back Big Daddy!" And the whole thing behind it was, I was tired of this guy I had grown up letting himself fade away and it bothered me. And that was my whole thing—to regenerate him. ❐

Jim "Jake" Jacobs

Woods and I were hanging out together when Ed offered him a job working on the Druid Princess. Woods accepted on the understanding that I get one also. Ed was skeptical so gave me a test. I had to clean up some flame-cut chassis brackets. Well, the files were all worn out so I had to work like heck to get the job done by the time Ed got back. He must have been happy, I got the job and it paid $2 which was more than I was earning at the gas station.

We worked on the Princess all that summer and then Woods went off to Vietnam and Ed said to me, "Well, what else can you do?" So I worked on a T-shirt display for the shows, signs, light fixtures and decal tables. I also repainted the Surfite.

The bike truck was also built at that time but I didn't have much involvement in it, other than hauling it around the show circuit, because Ed had gotten into the bike thing having decided that, "Bikes are where it's at and hot rods are dead." He began publishing *Choppers Magazine* and so I found myself doing the layout. I didn't know what I was doing but I was doing it. And it was fun.

Going to work with all those crazy characters: Dirt, Fuller, Woods and Williams, made for bizzare days. There was always something going on and mostly it was a comedy of errors.

Ed hung in on the bike thing, he really believed in it so the following year he began on the Candy Wagon and that kinda turned me onto trikes. Newton had done a drawing for the letters page of *Choppers* of a trike he called the "Mail Box" but it was originally called the "Booze Wagon" for a special three-wheeler edition of Roth's *California Choppers* magazine.

Meanwhile, over on some property my dad owned somebody had abandoned a Crossley and so I used the four-cylinder engine, trans and rear axle. I built my own frame and springer forks and bought some 8 1/2x15 American five-spokes but when it came to the body work I lost interest—I hate fiberglass—so I traded it to Roth who finished it off.

Above: Though the original Ed Newton artwork tagged this vehicle as the "Booze Wagon" its name was changed for its use on the letters page of Choppers Magazine. Jim "Jake" Jacobs, caught up in three-wheel fever, decided to build the trike using a Crossley engine but gave up when things got sticky. Ed finished it and this photo was taken at Movie World in 1975. *Photo courtesy of Tony Thacker.*

Rat Fink

Above: Ed fashioned the Rat Fink using a small cart frame and a Caddy electric-starter motor. Sadly, it scared little kids. Below: New York, 1966 and these kids look awestruck rather than frightened. Incidentally, the hand of the Rat Fink holds a small policeman. *Photos courtesy of Ed (above), and Doug Pizac and the Bob Hegge Collection (below).*

My mail order business for shirts & stuff kept growin' all thru the car buildin' years and out of all the characters I had on my shirts this one pesky little Rat Fink kept sellin' heads and shoulders above the rest. In the beginning it was just a rough drawing that I'd drawn on a napkin to explain what Mickey Mouse's father might look like. It was really crude. Then I trans-

ferred it to a refrigerator door. In all these drawings he had no hands & even though I drew it myself with no help from any of the designers I hung with I never really knew what his hands or backside looked like. It was only a picture y'know & I never even thought about him in 3-D so I decided to find out.

What better way than to make an RF car. Trouble was it was such a far out idea that I couldn't make it very big. I was gonna use a small Honda bike but needed four wheels so's it wouldn't tip over. The first Rat Fink I made was built on a small three wheel electric cart frame using a Cad' starter motor & 12-volt battery for propulsion. I was bored one day & mocked it up out of some spare plaster I had. When I got it done I got inside & chased some kids & it scared 'em so bad that I never got into it much after that.

The "Peace Rat" was supposed to keep kids from freakin' out—I decided to ride on top of it instead of

Top left: Early days with the Peace Rat. Notice the fingers are round the wrong way—unless you're English which is their equivalent to "the finger." Top right: The Rat resides now at the Southward Museum in New Zealand. Above: Ed's favorite flat four provided the power. ***Photos courtesy of Street Rodder (top left), Southward Museum Trust (top right), and the National Automobile Museum (above).***

inside it. It had all sorts o' bells and whistles and I had the exhaust comin' outta the mouth and to make it belch smoke I installed a small oil injector to squirt oil into the carb. Trouble was everyone that saw it thought that the V sign of the hands meant we were either dope-smokin' hippies or "Peaceniks" as they called them in those days so I was never really popular even though the V-Dub engine was a better motor than the Cad starter motor.

VW Trikes

Around '68, I was at some car shows back east & saw some guys that'd chopped the front end off'n their V-Dubs & welded Harley Davidson forks onto 'em, strictly for kicks! I went home & figured that the ideal speed & safety machine would be a VW built real low to the ground with some light forks up front. So I gets this really beat up bug, a 1957 with a 36 horse engine in it. I paid the dude a hunnert bucks for it. It was like takin' a family member when I dragged it off from his yard. His kids was cryin' & he had his head hangin' low & I felt like the villain Lash Larue when he ties the pretty maiden to the railroad tracks. I head for the shop & cut off the rear clip so's I can get the part of the frame that the engine's tied to.

I gets this 2 inch water pipe & makes this big humungous bend in it & weld it to the Bug frame. Then I welded a Honda 50 (it coulda been a 70) front fork onto this tube. Problem was that the tube frame & the forks weren't heavy enuff to keep the front end on the ground. Before I know it Dirty Doug starts pourin' plaster on the frame so's we could make a body for it. The plaster brought the frame to the ground with a thud and I told Dirt that as soon as we take the plaster back off it would do a wheelie again. "We'll cross that path when we get to it." was his reply.

So we worked almost day & night shapin' the plaster & gettin' a real low, low body—my buns was only 4 inches from the street. This was certain not to turn over when we got on the freeway off ramps. We covered the plaster with fiberglas & knocked out the plaster & mounted the glas body to the frame. Sure 'nuff! Whenever we sat in the seat the bike would sit "tough," but as soon as we got off the bike it would sorta do this wheelie. Y'see the motor was too heavy & it was like a teeter totter & the back end went down into this wheelie. So me & Dirt got some lead & melted it & poured it into this little hole I had drilled way up front by the forks in the frame. That did it! It sat straight and it only took 5 pounds to do it. Y'could still take the front tire & lift up on it & it was real light. So me 'n Dirt got the terrific idea to take this toad, later named "Hitler's Revenge",

Above: Ed's son Dennis checks the tire pressures of the first VW trike—Hitler's Revenge. Though he cut up a lot of Bugs, at least he salvaged and sold the small oval rear windows. After a lot of trips South o' the Border, Hitler's Revenge ended up in the Brucker's Movie World museum. *Photos courtesy of Ed (above) and Scenic Art Inc. (right).*

Left: Ed's son Charlie tries out what was to become, after a lot more work, the Wheeler Dealer. Below: Ed changed the name to Vitamin E and put it on display at Movie World. Bottom: Wheeler Dealer in its original guise. As Ed remembered it, he'd put a sign in his shop window advertising for a passenger and would get all kinds of volunteers. ***Photos courtesy of Ed (top), Street Rodder (middle), and Griff Borgeson (bottom).***

down to the river bed (we wuz three blocks from the L.A. cement riverbed), & test drove it.

Me & Dirt had chewed this big wad o' Wrigley's spearmint gum & used it to tie these little short lengths of yarn to the body. I figured if the front was gonna raise at top speeds then the yarn'd point upwards in the wind. So me & Dirt head for the riverbed with all these pieces of yarn danglin' from the body. I drove very gingerly 'cause I was afraid o' doin' a wheelie & dumpin' Dirt on his keester. He was white knucklin' it in the rear seat!. We got to the bottom of the river & get up to about 20 mph & she's steady as a rock. The yarns're all pointin' to the bottom & the steering is great. We go back & forth for about an hour & on each turn around I'd boost it up a few miles per hour. Steady! Handles well, & pretty soon we're peakin' out that poor 36 horse engine at like, 60 mph. I am a happy camper. Success—how sweet it is.

Later on we built some really big VW engines & drove 'em well over 100 mph & never had any trouble.

Fact is, I started ridin' with some guys from Las Vegas in 1969 & we went all over the country with those VW trikes.

I had seen the handwritin' on the wall when The Beatles came over. I knew the fat lady was about ready to sing. I had one trick up my sleeve though. I had met these two weirdoes at a show in New York. Woody & Sanchez. Woody was interested in buildin' bikes. So I made a deal with him to build VW trikes.

I had this habit o' takin' my trikes to the runs & givin' people rides on 'em. It all started 'cause I'm not a real heavy partyer or I never liked the games at the AMA runs so I'd tell kids I'd give 'em a ride for a quarter. Then I'd put 'em on back of the California Cruiser & try to scare the snot out of 'em. Well this scam backfired sorta 'cause soon I'd have these really long lines o' kids wantin' rides. Then the parents wanted rides too. I usually made me a hunnert bucks or so.

When I got the VW trikes I did the same thing only this time I'd get those Big Mamas with the Harley hats full of ride pins & invite 'em to go for a spin. They'd climb on & then I'd take 'em outta town a ways & do this bit where at 40 mph I'd do a 180. It'd scare the women but y'know what?! The followin' Monday after the run they'd be towin' their old man down to my shop & ordering a VW trike. Those trikes were dynamite. They handled & were very dependable. This bike thing lasted from '68 to '73.

In order to get enuff VW frames for the trikes, me & Woody use'ta go out at night & scout for bugs. I'd have one or two thousand bucks on me & a gun under the seat & we'd stop any bug driver & ask him if he wanted to sell it. In those days VW owners were very close to their V Dubs. It was like taking a family member from them. I can remember people cryin' as we drove off whenever we bought their favored Love Bug.

Above: The Tree Viper in its final guise is now part of the George Goodrich collection. Right: For his Secret Weapon, Ed lowered the center of gravity by using a lower, more radical seating position. It handled accordingly. *Photos courtesy of Tony Thacker (above), and Ed (right).*

Left: **Another in-construction shot of the Secret Weapon shows a silver metalflake paint job which as you can see in the shot below eventually gave way to army green. The sign says it's Big Daddy's prediction of what a Jeep for the United States Army should look like. Bodies cost just $570.00. *Photos courtesy of Ed (left), and Darryl (below).***

It was on a particularly cold night that me & Woody cruised over to East L.A. with the gun under the seat & me with my pocket full o' cash—we were goin' VW shoppin'. I spot this particularly cherry bug in a gas station with nobody in it so I pulls a you-ee & stops by the bug. Nobody around so I get out a paper bag & writes this note, "Are ya interested in sellin' this bug?" I stick the note under the windshield wiper & as I'm ready to get back into the car I hears the gas station attendant sayin', "Get away from that bug." Oh Boy. A Redneck. I says, "Is this your car? We wanna buy it & I left our phone number on the paper." "No, but I'm watchin' it for the owner." As he walks over to the cash register he adds, "Get away from it or I'll shoot ya!" I was so shocked by this guy's stupidity that I real quick & cool puts my right hand in my jacket & with my index finger outstretched (making believe I had a gun) said very coolly, "I wouldn't do that if I was you." He froze in his tracks. I jumped in the car & we jammed.

Me & Woody was crackin' up about the whole deal when all of a sudden all the cops in L.A. descended on us. The guy had called the cops! Of course, they found my gun under the seat & off to jail we went. What really blew the cops away was that I had the two grand on me so's I could bail me & Woody out without havin' to call a bail bondsman. Those few hours in jail was enuff for me. It cost me 350 clams for an attorney & I don't do that anymore! Then one day someone ripped off the entire shop & that was that. No more trikes.

All o' this wuz takin' it's toll on my family life of course. So it was no surprise when I found my suitcase on the lawn with a farewell note on it. 1970 was not my best year. I lived in a trailer in a parking lot while I got my new sea legs. From '70 to '75 I worked at "Movie World" in Buena Park, California. Jimmy Brucker had bought all o' my cars & he had about a zillion others that he was renting to the movie studios. He'd buy & sell cars too either at auctions at the museum or at private sales. You'd be amazed at the wheelin & dealin' that takes place in those back rooms.

With the "Secret Weapon" I figured that if I lowered the center of gravity with a more radical laydown position I could improve the safety factor that much more. The whole trouble was that it was too comfortable & I kept falling asleep at the wheel. It handled like a neat thing. At one of the "Cars of the Stars" car auctions, Jimmy dared me to run it through the auction. He said I could refuse any offer & I figured I'd get some mickey mouse bid of a thousand or so & I'd immediately refuse it. Surprise! Surprise! I got $3,500 for it which I immediately scooped up & took to the junkyard for some newer junk. It was one of the few I sold to a private party. Big Mistake.

Jimmy is a master dealer. I mean this guy can hardly tie his shoe laces but he can hack out a deal in quick order. I remember in 1970 when he was just about ready to open up Movie World he was hurtin' for movie stuff. He had plenty of cars used by the old movie stars like

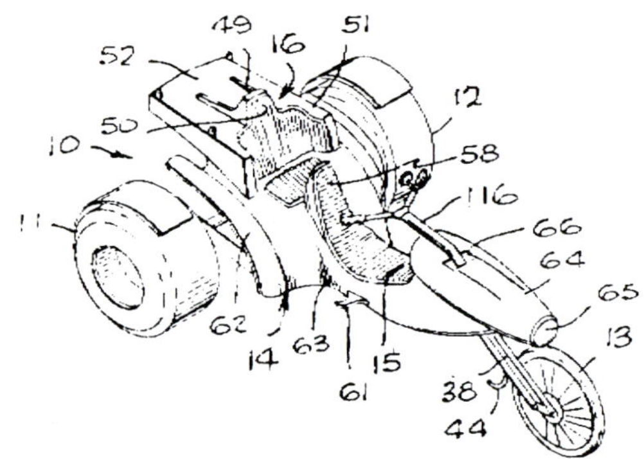

Clark Gable & Greta Garbo & Errol Flynn but no gutsy stuff. So he gets on the horn to MGM & Universal & makes deals to buy their old time props. Guess who gets to pick 'em up? Yup. Me. Can ya just see me climbin' around the ships used in Tora! Tora! & oversized champagne glasses? I had built my own tow rig & hauled cars & props for five years for Jimmy.

One of my other duties was to make signs for the cars. Jimmy's dad, Jim Senior, wanted the specs of the car on the signs. I told him if the car belonged to Clark Gable then a picture of Gable & Lombard oughta be by the car. I mean who cared what the wheelbase was? I did that over a lot o' flack from the old man.

To begin with, I didn't have a shop but Jimmy let me use the museum and that's where I started the experimental twin-tailed, Porsche-powered car fashioned after the famous Russian "Fox Bat" war plane. I had in the oven but it was too complicated. Besides, I finally saw that it resembled the Russian plane too much so I threw it away. I had better things to do than copy commie ideas. I took it apart & finished it as "Kolob".

I'd always had the hots for a Porsche sports car since they was first sold. But gads, they went for a fortune & sounded like a sewing machine. Your average run-o'-the-mill mechanic didn't know diddley about 'em so I just erased it from my mind. That is 'til I ended up with an engine thru some trade. Then I started huntin' for a Porsche box but no such luck. I found that the stock VW Bug tranny'd fit so bang! I was off to another super machine.

The VW trikes we'd been messin' with were a taste slow but the Porsche was perfect. I mounted the engine

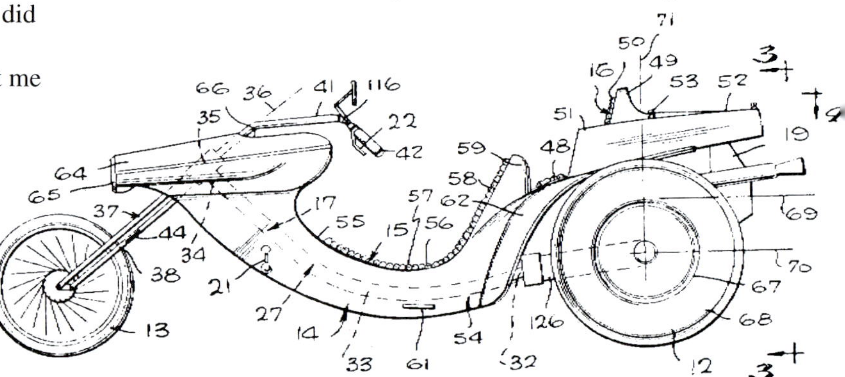

According to the United States Patent Office, Edward Roth was the inventor and patent holder of what was known as the American Beetle photographed here at Movie World in 1975. Right: Ed's motel while on the road is anywhere he decides to stop. Patent documents courtesy of *Ed. Photos courtesy of Tony Thacker (bottom left), and Griff Borgeson (right).*

midships so's I could get around corners without spinnin' out. I used Harley forks on the front & decided that an airfoil would make a good headrest. The six-cylinder mill hummed like a rocket & it was fast. I made the body outa fiberglas usin' my tried & true plaster system. I painted star constellations all over it & called it Kolob—the center of the universe.

There was a full-dress Harley rider hangin' out in Rialto by the name of "Tex" (don't ask why) who had the hots for fast trikes. He made me an offer I couldn't refuse and then covered the whole bike with nude women & told people he built it. Another lesson in sellin

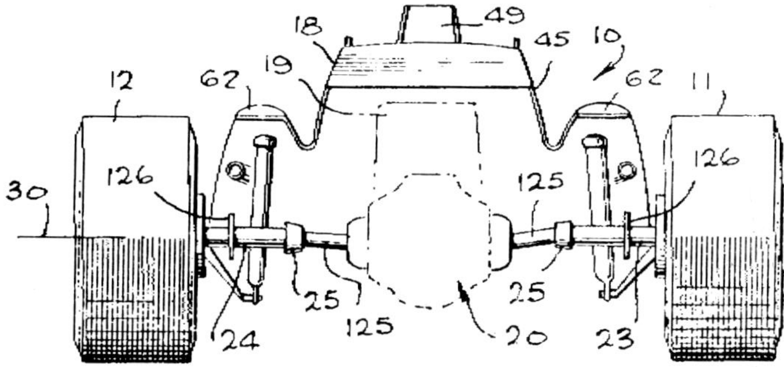

to private parties.

Von Dutch came to work at Movie World and took my place so I went to work at Knott's Berry Farm. I ain't said too much about Von Dutch yet. When he took my place at he'd just come from a two year stay in Arizona. He ended up there on account o' the cops was hasslin' him about his driver's license. He sorta had to give it up behind too many DUI tickets. So he just jammed to Arizona in about 1972. While he was there he said he was on, "The last road of the last town of the last state in the USA!". Dutch could'a been stuck out in the middle of the Sahara Desert & still get customers.

Dutch grew up in Watts which's now South Central L.A. He's told me how he didn't mingle even at an early age. He even ate dinner by himself. He made his own zip guns to protect himself from the notorious gangs that roamed the streets in them daze. He'd always get a passin' grade 'cause the teachers'd stick 'im in back of the room & back there he'd sketch everything from erasers to thumb tacks. Later on he went to work at a bike shop as a mechanic & when this one particular bike came in for a paint job he did it using his dad's brushes (his dad was a brilliant painter) to finish off the pinstriping on the fender & he never twisted a wrench after that. He told me in them days he'd use just a regular sign paintin' brush for the pinstriping.

Anyway, me & Dutch got along pretty good over the past 25 years. He was the best comedian I'd ever run across, but he couldn't dig on people that wanted to exploit him. He had a gun under all his benches & everywhere in the shop. He'd shoot at people & chase 'em away 'specially after he went out to the Brucker ranch in Santa Paula. The ranch was surrounded by this big fence & I had ta ring this big cowbell to get his attention in the trailer that Brucker built for him. We were on the same wavelength 'cause we both built neat stuff.

I heard about Dutch while I was still in the Air Force ('51 to '55). He'd just come from the bike shop & took up with the likes of Barris Kustoms on Atlantic Boulevard in Lynwood. Y'see Barris could get him all the high dollar work he could handle. Besides, Barris needed someone to cover up the zits in his paint jobs & Dutch was an expert at that. Then he moved to Competition Motors in Hollywood where he started hittin' the sauce a bit. Then in about '69 he moved to Calabassas where the Movie Star retirement home is. That's where I first run into him in person.

He was still married to Sheila & he did all his stuff in the small garage he had out back. I went out there & picked up a VW-powered XA Harley he'd built. I knew he needed bucks so I always paid his price. He always knew what his stuff was worth & he wouldn't back down. In fact if he got too much flack he'd raise the price. At about this same time I made a deal with Canning's Car Shows to bring in Dutch as a guest artist. Now Dutch had this real unique way o' gettin' people upset in a big way by tellin' 'em the truth. I got him to come over on the Saturday of the show. There wuz a lotta people waitin' there for him. He was gonna draw this picture on a sheet o' Masonite I'd cut up & mounted on a easel.

So he's drawin' away & rantin' & ravin' about this

& that & the picture was takin' shape. It was this back side of the moon with a bunch of 2 by 4's nailed to it. He said since no one saw the back of the moon, "It could be a prop held together with nails. Who knew?" Then, when he was almost done he whips around & starts talkin' about peoples false values. He says, "Now if a horse takes a dump in a field people'd say, 'Man, How gross! It smells! Cover it up'. But if the horse took a dump & it was gold plated people'd scoop it up & love it." I thought Newt was gonna crawl outta his skivvies. Dutch used a lotta other words & the story was quite a bit longer.

Then he went to Arizona for a coupla years & corrupted some dudes over there including "Slow" & "Butcher." When he came back from AZ he stopped at the museum to see me & he was drivin' this front drive Olds & since there was no drive shaft he'd installed this real ornate highly engraved brass tube where the drive shaft tunnel would normally be so's he could dump his beer cans out without anyone seein' where they was comin' from.

After he took over at Movie World, I used to stop in once a week or so to see what he was up to. For the next five years ('75 to '80) he created some of the most outrageous displays ever built. Then he moved to the ranch in Santa Paula. That's where he made the best stuff in his life. He was alone out there & had plenty o' time to develop his skills. He'd always call me when he made some neat stuff. I'd go out & pay his price & sell it for charity at the next Rat Fink party. He was always at the parties. Even though he consumed a case of Miller's beer every day, he liked Coors better but he said he bought Millers 'cause he liked the way the "M" was made. On days of the party I'd have to drive out to the ranch & pick him up & he'd never have a drink before I got there. Since Dutch started the whole car decorating craze all the people at the party would treat him like the king that he was.

One time I remember goin' out there & he'd built these great guns & we went out to his private shooting range where I got to shoot at all the strange danglies he'd put up for his targets. His guns were neat. Another time he had this pressure-powered BB machine gun. When I asked him if it was accurate he whips around & shoots out the nearest light bulb with a single burst of BB's. Then there's the time I asked him to take one o' his cap & ball pistols & shoot a hole thru a trash can so's we could auction it off at a RF party. He did that for me! He dug on how far bullets would penetrate a board or a piece of steel. BB's would go further than regular bullets.

Then here's my last story (I hope). Dutch had this bad bad dog named "Challenger". He was ugly & old. Dutch'd throw him a can o' dog food & this dog was so

Below: Yet another VW-powered trike this one tagged the Panzer Trike which is now in the collection of George Goodrich. *Photo courtesy of Tony Thacker.*

Top left: The Porsche six-cylinder engine from this ill-fated rocket-type vehicle went to power the mid-engined Kolob around 1975. Ed eventually sold the vehicle and we think it still exists in the So-Cal area. ***Photos courtesy of Ed (top left and below), and George Goodrich (above).***

bad he'd chew it open & eat it. His teeth was almost not there. So one day I brings this Pizza to the ranch for Dutch figurin' he'd dig it. He opens the box & tossed it to the dog & said, "I don't need food, I get all my vitamins from the booze!"

Dutch started goin' to the Santa Paula airport every mornin' 'cause he wanted to build a VW Thing airplane. So he studied the planes. He was gonna make the wings outta electrical conduit. He'd sit around with these old war dogs & he even got to look a little like 'em. At that time he'd drink coffee but on the way home from the airport he'd stop & get the beer to fill his coffee mug.

The bike he rode had this giant sidehack on it. I asked him why he made the hack longer than the bike & he told me, "In case I ever have to haul a bike back here it'll fit in the hack." He died of liver cancer September 19, 1992, & he always told me, "I invented modern pinstriping & when I die it'll die with me." It's true. There's a little Von Dutch in each and everyone of us.

Rubber Ducky

About 1972 I bought this new pad in La Mirada, California, (another wife too!) down the street from the Farm where I upset the people of this high falutin' neighborhood with my overspray and fiberglass dust but finally had a garage again. Trouble is the cops made me close shop at 10pm to stop the complaints. I merely shut the garage door and did quiet work.

Why "Rubber Ducky"? It all started with those crazy little Honda cars that first hit the US around 1970—y'know, the ones that look like a giant hockey puck. They cost a buck a pound & weighed 900lbs. They were at the dock in Long Beach waitin' for the big wigs in Sacto to pass on their legality. It only had a 600cc engine.The honchos were tryin' to decide if this lil' car was able to pass the five-mph test.

Soon they hit the Honda dealers & I felt if Mr. Honda had built it then it must be bulletproof. I bought one. Not to drive but to destroy! I mean I wanted it to blow apart! In fact I had two. One beater & another one I chopped into a pickup truck. Get this. The engine was air cooled & had no guts but ran forever on a gallon o' gas. I couldn't destroy it. I rolled it & got run into & all's I'd do is crawl out the door that was pointin' to the top & lift it back on all fours & keep boogyin'.

In my brain I figured if I took this really great

Far left: Unibody construction gave strength and a 26-gallon gas tank in the Kevlar-bodied, 600cc Honda-powered Rubber Ducky now in the Southward Museum in New Zealand. Left and below: Plenty of plaster seemed to be needed for the beaky automatic, VW Type 4-powered Great Speckled Bird. *Photos courtesy of Street Rodder (page 114), and Ed (page 115).*

engine (even Dirty Doug got one) & built a trike outa it I'd have me a real winner. It had the tranny & engine in one piece. And simple too! It took me about three months to slap it together. Weighed 325 lbs. The trick was I'd made the body a 26-gallon gas tank. I mean the whole body held gas. I did that 'cause I needed the range to beat the bikes I'd ride with 'cause top speed was like, 60mph.

I got it done & decided to tool around town for a test drive. Bingo! Pay dirt! The cops stop me. When I know I gotta winner is when the fuzz gets nervous. "Ya got a license for that go-cart bud?" says the cop. Now I know I'd miniaturized the bike by usin' all small tires & stuff but my big bod (280 at 6'3") made the thing look really small. After showin' my papers he cut me loose. I think he was sorta upset! Tough!

Now for the real test. L.A. to Salt Lake to Frisco to L.A. My son Dennis wanted to go along. Same with Bob Curr from the Knott's sign shop. Dennis had a Honda & Bob had a Beamer R-600. It was summertime & hot. Like I predicted they'd stop for gas & I didn't. Except for Dennis freakin' out with dehydration & rollin' in the sand like a walrus from all the Pepsis he drank (Oh Dad! They'll cool me down! Sure!) The trip went well. I was gettin' 85 miles per gallon & finally refueled in Battle Mountain, Nevada, leavin' the gas station attendant speechless with the amount o' gas I put into the "Duck".

I had left LA with the clothes on my back & Dennis was carryin' my sleepin' bag. I had on cowboy boots & I always wear polyester pants & dress shirts so in case I go to jail I go in style. Those loose pants kept blowin' up around my knees exposin' most of my legs & I never paid much attention to it. The second day I felt the sunburn but kept jammin'. Third day, blisters. When I got back to L.A. the fourth day my legs were a big red mass of blisters. It took skin & hair & all. To this day, I've got no hair on my legs. That's the price ya pay for tryin' to act like a big shot! I believe the Duck's in some private collector's hands now.

Next I worked on the "Great Speckled Bird". An automatic tranny made this one a more pleasurable machine to ride. My new wife didn't like it even though I installed a five-gallon water tank & a windshield wiper motor so's we could squirt our hot bodies while cruisin' the desert. It worked! In fact that little invention cooled us down so much that other guys useta follow us so's they could get part of the overspray.

I can still remember a trip to Ensenada, Mexico, where all the dudes were tryin' to show off their Sportsters & other type bikes on the beach. All they could do was spin the beach sand with their back tires & get into a rut, & get nowhere. I just cruised over the sand like it was glass & made no ruts or spun no dirt. Those guys on the Sportsters were fit to be tied.

Asphalt Angel & the Pink Bazooka

Then one day in '86 I decided to make a video of buildin' a "Cobra" trike so I buy this new video camera & start recording all the building of the "Asphalt Angel". (I still have the video available). It was okay especially since I was tossed outta all the rod runs except Goodguys'. The publicity was worth it. I needed a pickup truck in '88 to move from L.A. to Utah so I traded Holmes Mini Truck Salvage the Angle a really nifty Peterbilt mini truck. Well as far as anyone knows it's a Peterbilt. I put a "Peterbilt" sign on the tailgate to disguise the fact that it's a Mazda 2000. Y'see it's not hip to drive Japanese stuff 'specially down south. If you've ever stopped at on o' those "good old boy" cafes in Texas with all the Ford pickups it's like a crime to even drive a Chevy pickup much less a Mazda. I'm not proud of havin' to use Japanese stuff. It's really downgrading to have to pose my truck as an American product, but there's some o' them cowboys that start makin' fun of overseas stuff. My only defense is that when I'm on a tight show schedule I've gotta be there on time! No excuses! That Mazda delivers. Never late & never any problems (Unless ya count havin' to rewire the headlights on the fly as "Major"). When I had Fords I had problems. The Mazda has well over 120K on it & I gotta be in Boston for the New England Nats next week. No problem!

Far left: Still in La Mirada, Ed decided to make a how-to video and built the Asphalt Angel using a V6 Buick engine. He rode all the way from L.A. to St. Paul, Minnesota, to go to the Street Rod Nationals but they turned him away. Above: Ever inventive and up to date, Ed tried solar power to back up the Honda 50-powered, Kevlar-bodied Pink Bazooka. It's now in the George Goodrich collection. *Photo courtesy of David Fetherston (page 116), Tony Thacker (color), and Street Rodder (B/W).*

Globe Hopper

One night in 1987 I had this dream that I was travellin' across Russia in one o' my vehicles & when I woke up I was sorry I had the dream 'cause it's a lotta trouble gettin' across Russia or China. I pleaded with Father to take back the dream but it didn't go back so I planned this super "outback" trike to go long distances. Luckily I was able to go to Alaska instead of Russia. Phew!

I'd been ridin' trikes for fifteen years when I decided to make the "Globe Hopper"—named by Robert Williams of course—a trike with a big gas tank & a trouble-free VW mill. It had to have automatic transmission so's I could cruise instead of drive. I settled for a, 1800cc 914 sedan engine and auto. I remember buyin' the whole car for twenty five bucks. It had fuel injection but I ripped that off the first night amidst screams of, "Quite out there." from the neighbors. I got two SU carbs from Rivera engineering in Whiter & made some real tough air filters outa inline filter material. I knew I was gonna run into bad roads (Bad? That ain't the half of it!) I made me an extra-heavy frame & steering & a two-seat bod to store extra grub and clothes.

I told everybody I knew I was going to Alaska then across Canada to the Canadian Nats in Kitchner. I even called Boyd Coddington & asked if he wanted to take one o' his fancy rigs & go along. No deal. They all had excuses.

Come the night of departure, it was summer time and warm as toast but boy was I in for a surprise. I didn't have many extra clothes. It was like a trip around the block. Oh, I had extra blankets & a sleepin' bag so I went to the drive-in where all the local rods hung out & I told 'em I was goin' to Alaska. They all cracked up & invited me me for pie & ice cream. I ate the treat & said, "Thanks guys, I'm still off to Alaska." And off I went. I had a bag full of clothes & stuff & had a Rat Fink mask draped over it so' the kids'd get a kick outa it while I was drivin' along. I was gonna remove it after Fresno but kept it the whole trip. Of I went into the sunset with the dudes all crackin' up.

Everything went smooth thru California & Oregon & Washington but when I got to the Canadian border it started to get a bit chilly & rainy. I had no windshield but by "peepholin'" (makin' a small circular hole in the face mask with my index finger to peep thru) I got way into Canada & Alaska. I gotta mention when the rain started this dude pulls me over on the freeway & & offered a prayer right there on the freeway. Thanks dude!

The further north I got the colder the weather & the rougher the roads. Norm Grabowski had warned

---

Having been turned away from the Street Rod Nationals, Ed tried another tack and used a body something akin to a '34 Ford Roadster for the Globe Hopper which he intended to drive across Russia but instead took a wrong turn to Alaska. *Photos courtesy of David Fetherston (above), Terry Thompson (left), Lou Martineau (top right), and George Goodrich (right).*

Lou Martineau

In June 1987 my brother and I, our brother-in-law, and a friend, were returning to North Dakota from a fishing trip to Lake of the Woods, Canada.

We were traveling west on Canada Highway One when our motorhome broke down. We were trying to fix it when a man driving a small, three-wheel car went by. The little car slowed down and made a U-turn, stopping behind our motorhome.

This rather large, bearded man got out of the vehicle and asked if we could use some help. As it turned out, we were very glad of his assistance and after the motorhome was fixed I asked if I could take pictures of him and his car. He was happy to oblige. The car looked like a racy, miniature hot rod painted gray with flames and the name "Big Daddy" on the side. The owner told us this was a one-of-a-kind creation in which he had traveled from southern California, up the coast into south Alaska, then southwest across Canada.

We didn't realize until a few weeks later, when we were looking at my trip photos, that it was in fact Ed "Big Daddy" Roth who had stopped and helped us on the road and I share this story with you in tribute to someone I got to know by chance—Ed "Big Daddy' Roth. ❐

IT WAS COOLD! ALASKA '87
"GLOBE HOPPER" RAT FINK AND ME!
ED "BIG DADD
ROTH
1990

Above: After the Alaskan saga, Ed in his infinite wisdom, sets off once again for the Street Rod Nationals only to break down and abandon it in Oklahoma from whence he and George Goodrich eventually retrieved it. It is now part of George's collection. *Photos courtesy of Ed.*

me about the rough roads 'cause the sun'd only come up part way on the horizon & so with the big trees the sun wasn't able to melt the water on the dirt roads so they turn to mush & stayed that way. No place for a California Cruiser.

I can remember when I hit the first mush road. I was near a town called Stewart. It just went from asphalt to mush. Bloop! Just like that. I figured, "Well, it'll turn back into asphalt pretty soon. This must be a detour." Well, the real story my friend was that it lasted way up into the Yukon & then to Dawson Creek where I picked up the asphalt again. That was about 600 miles and a zillion mosquitoes later.

One cop stopped me & axed me where my windscreen was. Gas stations? I'm glad I had this big tank with a lotta reserve. About 200 miles between stations & cities. Lotsa moose & critters hardly no traffic except these big loggin' trucks that splattered mud all over everything. Luckily I never had a windshield. Whenever the loggin' trucks splattered me I'd take my glove & wipe off my shade with one clean swoop & keep goin'. I felt sorry fer the cars. I stopped at one glacier & I was so cold I couldn't move but I waited 'til a car came by to have 'em take me & "Finky's" picture by the glacier. If you're plannin' a run to Alaska, see me first.

So I get back home in time to get ready for the Street Rod Nats in Louisville, & to do a TV special on the "Gidget" show. I had a new flame paint job & felt indestructible. I knew the NSRA was gonna kick me out when I got there but I always had a blast at the stops renewing old friendships.

I'd conned 200 free pizza tickets from Domino's & provided pizza for the gang at every stop. Flagstaff, Albuquerque, Amarillo, Oklahoma City & then it happened. The temp gauge went outa sight. I kept good tabs on the oil level but one of the air vents of the air-cooled bug engine had flapped shut & made the whole engine hot. Ka-pow! That was all she wrote. I pushed it in a field & Greyhounded home. This field was pretty well hidden but had this humongeous horse in it. He was my pal but he didn't like people.

I came home dejected & was gonna just write the whole thing off as a bad scene. I'd always told my wife if I broke the bike, specially on the Alaska gig, I'd burn it and hitch a ride home. So I'm broodin' away when I get this call from George Goodrich who has a strange fascination with my stuff. I tell 'im about the "Hopper" bein' in this field in Oklahoma & he says, "How much ya want for it?" We made the deal but George insisted that I go along to pick it up. Well, nobody messed with it while I was gone & George towed it home & got it runnin' again & takes it to cars shows still. I'm stoked that this story had a happy endin'.

Dennis "DR" Roth

All my brothers will probably tell you that I was my dad's favorite but the only reason that I may have been was because he told me, "If you don't snivle then you can hang out with me." And I was just glad to go to Texas on a V8 trike and have fun and go hang out and work at the car shows but I guess my brothers took that as I was his favorite.

The first thing I can remember was I sanded the seats for the Mysterion. There was two seats but there was only one seat in the car. He put legs on the other one and we had it at our house and the head rest, which was like a bullet, had an eyeball painted where you put your head. It was one of these big swooped-out things with your legs way out.

When I was about 14 or 15 I was on work experience in high school and I'd come home and flip the little Volkswagen trike bodies over and I'd grind and I'd laminate in the gas tank and the mounts—I had a little jig.

Dad had a little jig that he made called Mr. Big that did everything and Mr Big had a tank head with a little bolt coming off and this little propeller on top and if Woody or Big Daddy had any questions they would twirl his proppeller and if it was one way it was YES and if the other way it was NO. And they used to go by that too.

I remember him going to Laidlaw's to buy a new Sportster and when he got home mum says, "That's a death trap." Well the next time my mom sees it it had Death Trap lettered on the side of it. She did not like it.

When Mickey Thompson or Ansen wheels or any of these people wanted shirts, Big Daddy he'd go, "Okay Mickey, I'll make your shirts but trade me for fuel." So we'd go down and give Mickey his shirts and get nitro from him. Ansen, wanted shirts so dad goes, "Hey, Newt will design 'em and then we'll trade 'em for wheels."

The post office at one time refused to pick up our mail because it got way too large so dad bought this truck and he painted it red and on the side it said, "Why Keep Your Business a Secret." He also had this F-100 panel and I told dad, "If we cut a hole in the roof we'd could haul a lot more trash." So you know what he told me? "Go get the torch and get the hose and wet the trash down inside."

Somewhere someone is trying to restore that bitchin' panel, I mean it was bitchin' there was no dents in it, nothin' and the old man just cut a hole in the roof. He goes, "Okay, just let the roof fall down on the trash." And we just kept pilin' on more trash. That was our trash truck for a long time until they started pickin' up our trash again and that was years later—they said there was too much paint and goop.

By the time Dirt left and I started doing the plaster and fiberglass for dad that was probably around the time of Captain Pepi's. I also worked on the Druid Princess but that was mostly plywood so there wasn't much to do on that but then came all he trikes. That was my job, mix, mix, mix. Dad shaped it all and I just mixed the plaster and the he'd make it out of sticks and chicken wire and then he'd shape it. It could take a month or a year to do that depending on how many times he chopped it up and made it look different but once we started with the fiberglass it was a monolithic deal so we just had to start and go. He showed me how to do that. After a while he showed me how to make a hot batch or a slow batch or I'd know because I'd know if he was doing flat work or going up the sides or on the top because once you start glassin' it you just gotta keep going' until the whole thing's done. And I remember mixin' in upside down German helmets, gallons of the stuff. Because dad would get it in 50 gallon drums and I'd just go over there and fill up helmets, put the MEK in it, mix it, give it to him, cut cloth, and keep going—it was a sad deal.

It was pretty smooth when it was plaster. First he'd put down the glass, and then he'd put down the mat over that so he could squeegee it all out and then let it dry and then somehow chip all the stuff out. But dad told me when I was six that he made the Beatnik Bandit totally out of plaster with no vemiculite in it and that him and Dirty Doug actually had to shape the hard plaster with grinders and then they painted it because Revell wanted to come and measure it early so it wasn't even fiberglass when Revell came to make the model of the body, they just came and measured that plaster and they thought they were looking at the car and dad had to paint it like it was eventually going to be painted. They didn't know what they were looking at. And then he'd chip all that stuff out from the back.

L,A, Zoom

In 1988 I got an Acura & pulled the engine & guts out of it. I saved every little wire & instrument & put it into the L.A. Zoom frame 'cause I thought it was time to prove that computerized engines could do the job. I was also messin' with this new Kevlar glas that Rutan used on the "Voyager" airplane. It is stronger & lighter than regular glas (more expensive too!) In fact they use it for bulletproof vests.

When I got it almost all the way done I trailered it to Utah & was gonna finish it here where I can still make a lotta racket & overspray. The motto around here is, "If ya don't mess with my cattle (or sheep) I won't mess with your cars." I started inquiring around for someone that could hook up the computer. No one could. So it's sittin' in the shop waitin' for that certain genius that can crank

Above right: Thom Taylor's original artwork for the ill-fated L.A. Zoom which was to have been Honda-powered. Sadly, Ed could not find anyone capable of getting the computer-controlled engine to work and so sold the project to Terry Holmes of Holmes Mini Truckin' where it still sits. ***Artwork courtesy of George Goodrich. Photos courtesy of Ed (top left and bottom left), Street Rodder (left middle).***

it up for me. Recently I tried to unravel a Honda 150 cc wiring diagram. No way! Von Dutch use'ta tell me, "Whenever ya tear apart a Jap machine & poke all the little black boxes with a sharp screwdriver all ya find is a bunch of "dirt". It's true!

I was livin' this really normal-type life with wife numero deux when it all came to a screechin' halt one day when I came home from the Boston car show. My bags was packed & layin' on the lawn. I was history. The judge says move! What? Move? Yup, you're out!

I was depressed. Not about the wife but about the cars I was buildin' in my shop at the time. So I jams out to the airport to see my friend Gary who had been in 'xactly this same predicament a little while before me. I was just gonna dump the whole mess in the snow in some field around here. But Gary says, "No problem. We'll take it over to the college." Ya sure I'm thinkin'. "Besides that, there's all o' these big machines in there that ya can use," says buddy Gary.

Turns out he's right. They've got this "Incubation Program" where they let citizens use the college facilities to do things that bring money into the county. So I build cars, or whatever else I need to do: shirts, etc.

The judge made us sell the big house & shop & split the money & I moved into this adobe house & the roof use ta leak & the adobe got wet & melted right before my very eyes. It's a long story, but the judge is happy & so's the ex & so'm I.

Left: Ed has never had too many projects that have gone uncompleted. There was the Porsche-powered rocket thing that became Kolob, the L.A. Zoom and this steam-powered car that is seen here at Movie World Cars of the Stars where Ed was building the body on a VW floor pan. Sadly the engine, which he had converted to run on steam some time earlier, was stolen. ***Photo courtesy of Ed.***

Mini Wagons

After I recuperated from the divorce I decided to build some small projects to have some real fun with. I'd seen these motorized wagons up in Seattle at a "Goodguys" run & asked him how much he wanted for it—I needed the wagon to get from my T-shirt booth to the hot do stand and restrooms. When he told me $3,000 bucks I almost passed out be he added, "I ain't got time 'cause I'm backordered for six months." Get serious. For that kinda bread I could build half a dozen of 'em.

I went home & got all the parts & used some valve springs & pipe nipples to make the miniaturized Mysterion front suspension. An American Flyer wagon from Toys R Us & a Briggs & Stratton engine & I was all set. Zip! Zip! Bang! Bang! & I had me a real fun machine named the "Conastoga Star" by Robert Williams.

I drove it in parades dressed up like a cowboy & I'd blast the kids with this giant water gun. I drove it at an air show down the runway & let super light airplanes bomb me with flower bombs. Then I took it to car shows

Top left: Ed took down-sizing to a new level when he got into building these funky mini wagons at Snow College. Shown is the wheelbarrow-based Yankee Blitz. Below left: The Booger Buggy is now owned by George Goodrich. Below: Ed aboard the Conistoga Star. Right: Ed turns to Sour Grapes with his Honda-powered lawn tractor. Below right: The five-wheeled Fink Mobile powered by a Honda Elite. ***Photos courtesy of Ed (top left, right and below right), Tony Thacker (Bottom left), and David Fetherston (below).***

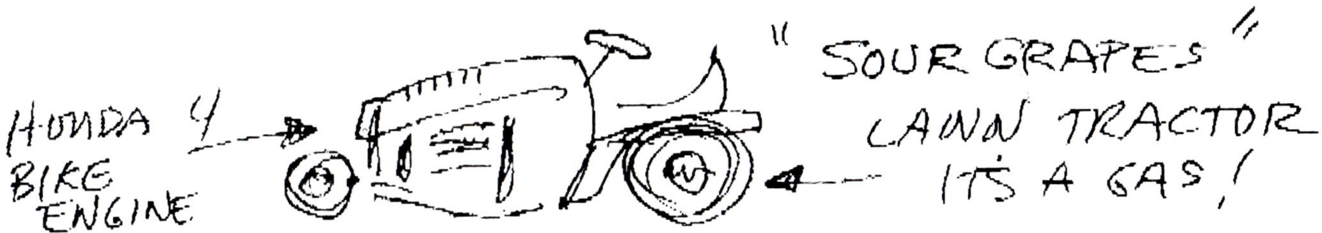

where the reporters preferred it to the bigger show cars 'cause I'd tool up & down the aisles with it & they picked up on it & made a big deal out of it in the local newspapers.

I finally donated it to the National Auto Museum in Reno, Nevada, where it is on permanent display right alongside the Beatnik Bandit. Meanwhile, I'd been searchin' for a wheelbarrow to do the same thing with. Thus was born "Yankee Blitz" again named by Robert Williams.

Top: Over the years, despite liking sleepers, Ed has had some pretty wildly decorated drivers such as this bug used in the mid-'70s. This was sold to buy the '76 Honda above which BDR used until it was sold to a Japanese buyer in 1986. Left: This little Honda saw work as a trike body deliverer until it was purchased by George Goodrich. *Photos courtesy of Ed.*

Beatnik Bomber

Recently I sold my Rat Fink portion of the business to the Moon Eyes company in Santa Fe Springs, California. I'm ready to retire & make that movie or collect that social security. But wait a minute. Is that where it's all at? No way Jose. I've got plans & as I write this book I've got the "Beatnik Bomber" (Beatnik Bandit II) almost halfway done. I love to work on cars & when the final curtain comes down I wanna be wailin' on a car.

In the back of my head are enough projects to last half way thru the millennium. My next great challenge is to build some small water tight hot rod trailers that be "Lookin' good", & to settle down (Oh no. Not that) My personal life has taken a great setback because the crazy hours & the fiberglass dust & time away from the home.

When I visited Ed at Snow College, he was working away at Beatnik Bomber a 350 Chevy-powered tribute to the original Bandit and something with which he could enter those wild burn out contests they have. *Photo courtesy of Tony Thacker.*

The future (Yech!) holds many wonderful & marvelous things for hot rodding. Computers (double yech!) will open the doors to more efficient ignition & injection systems. Virtual reality will allow us to build cars before they're built so we can see what it looks like before we get out the plaster or clay for the model. I hope that in some way whenever kids look at my stuff they can help their own careers along by saying to themselves, "Gee, if "Big Daddy" Roth could do it, so can I."

Index